Analog Circuits and Signal Processing

Series Editors

Mohammed Ismail
Mohamad Sawan

For further volumes:
http://www.springer.com/series/7381

Tom Van Breussegem
Michiel Steyaert

CMOS Integrated Capacitive DC–DC Converters

Springer

Tom Van Breussegem
K.U.Leuven ESAT-MICAS
Kasteelpark Arenberg 10 bus 2443
3001
Leuven
Belgium

Michiel Steyaert
Department of Elektrotechniek
K.U.Leuven
Kardinaal Mercierlaan 94
3001
Heverlee
Belgium

ISBN 978-1-4614-4279-0 ISBN 978-1-4614-4280-6 (eBook)
DOI 10.1007/978-1-4614-4280-6
Springer New York Heidelberg Dordrecht London

Library of Congress Control Number: 2012940719

Printed on acid-free paper

Springer is part of Springer Science+Business Media (www.springer.com)

Preface

Monolithic integration is the paramount trend in both consumer and industrial electronics. In only five decades computers turned from room-filling machines into devices that fit in the palm of our hand. But a computer is only a single example of the large number of electronic devices that surround us in everyday life. The monolithic integration of electronic circuits—i.e. radio-transceivers data-converters complete digital signal-processing systems—has led to a tremendous increase in portability of the state-of-the-art electronic appliances.

But a single building block remains difficult to be integrated in a monolithic electronic system: the switched-mode DC–DC converter. The DC–DC converter provides an interface between the power source—whether it is a battery, a high-voltage DC bus or a loosely regulated supply—and the different voltage rails required in an electronic system. In most cases the switched-mode DC–DC converter is implemented by means of a separate chip, with discrete-type components or a monolithically integrated linear regulator is used instead. Each of these solutions leads to either a bulky, expensive or low power-efficiency solution. This is unacceptable in times where power savings and cost reduction is the governing social paradigm.

Switched-mode DC–DC converters are roughly divided into two categories: the inductive type and the capacitive type. The first using both an inductor and a capacitor to convert the input voltage into a regulated output voltage, the latter using nothing but capacitors to achieve this. In theory inductive-type DC–DC converters provide a lossless DC–DC conversion for a continuous input-output voltage range. Capacitive DC–DC converters fail to meet this expectation. And therefore inductive-type of DC–DC converters are the dominant type of DC–DC conversion apparatus in both commercial and industrial prototypes. For a long time inductive-type DC–DC converters were thought to maintain their superiority even for monolithically integrated prototypes. But in an integrated case the inductive converters are cut short by the poor quality of the integrated inductors, the key-components in the design. Therefore the intuitive preference for inductive converters does not hold anymore. Moreover, integrated capacitors—crucial for the operation of the capacitive converters—are native devices in CMOS technology and can be constructed at high quality. Therefore, despite their obvious limitations, capacitive DC–DC converters are viable alternatives for the inductive counterparts. But the adoption of monolithic capacitive DC–DC converters requires an extensive analysis of the conversion characteristics. This book describes the background required for designing a fully integrated DC–DC converter in CMOS and provides a detailed discussion of a number of CMOS prototypes.

Contents

Abbreviations, Symbols and Quantities

Abbreviations

ABB	Adaptive body biasing
AVS	Adaptive voltage scaling
AC	Alternating current
CMOS	Complementary metal-oxide-semiconductor
DC	Direct current
DDSM	Deep deep sub micron
DUT	Device under test
DVS	Dynamic voltage scaling
EEF	Effciency enhancement factor
ESD	Electro static discharge
FF	Feed-forward
FSL	Fast switching limit
FBB	Forward body biasing
FOM	Figure of merit
GBW	Gain-bandwidth
IC	Integrated circuit
ITRS	International technology roadmap for semiconductors
IVCR	Ideal voltage conversion ratio
LDO	Low drop out regulator
LUT	Look up table
MIM	Metal-insulator-metal, type of capacitor
MOM	Metal-oxide-metal, type of capacitor
N/PMOS	n-type/p-type MOS transistor
N	Voltage conversion ratio
OLG	Open loop gain
OIB	Output impedance balancing
PCB	Printed circuit board
POL	Point of load

PMU	Power management unit
PWM	Pulse width modulation
PFM	Pulse frequency modulation
RF	Radio frequent
RMS	Root mean square
RBB	Reverse body biasing
SOC	System on chip
SIP	System in a package
SBB	Static body biasing
SMPS	Switched mode power supply
SSL	Slow switching limit
TR	Topology reconfiguration
TAA	Time averaging approach
VCR	Voltage conversion ratio
VRM	Voltage regulator module
VDS	Voltage domain stacking
VSS	Variable structure system

Symbols and Quantities

A	Area
C	Capacitance of a structure
C_{out}	Output capacitor
C_{fly}	Flying capacitor
C_g	Equivalent gate capacitance of a MOST
C_{sq}	Capacitance per m^2
C_{xy}	Capacitance between node x and y
I_{in}	Input current
I_{out}	Output current
G_{tot}	Total conductance of the switches
G_j	Conductance of switche with index j
L	Length of a transistor
L_{eff}	Effective length of a transistor
L_{min}	Minimum length of a transistor
N	Voltage conversion ratio
P_{load}	Power delivered to the load
P_{in}	Input Power
P_{out}	Output Power
P_{loss}	Power loss
P_{gate}	Power loss due to charging the gate capacitance of a transistor
P_{par}	Power loss due to charging the fringing capacitance of a capacitor
$P_{R\text{out}}$	Power loss due to the output impedance of a capacitive DC–DC converter
R	Set of operation points of a DC–DC converter

R_x	Operation point of a DC-DC converter
R_{out}	Output impedance
R_{SSL}	Slow switching limit output impedance
R_{FSL}	Fast switching limit output impedance
V_{bs}	Bulk-source voltage of a transistor
V_C	Voltage across a capacitor
$V_{C,0}$	Initial voltage across a capacitor at t =0
V_{dd}	Voltage of a supply
V_{ds}	Drain-source voltage of a transistor
V_{gs}	Gate-source voltage of a transistor
V_{in}	Input voltage
V_{out}	Output voltage
$V_{\text{out, iny}}$	Output voltage considering an input voltage V_{iny}
V_{th}	Threshold voltage of a transistor
W	Width of a transistor
a_c^i	Capacitor charge flow vector in state i
a_s^i	Switch charge flow vector in state i
d	Distance between two structures
f^{sw}	Switching frequency
q_{out}^i	Charge flow element of the output in state i
q_{in}^i	Charge flow element of the input in state i
q_x^i	Charge flow element of component x input in state i
α	Percentage of fringe/parasitic capacitance of a capacitor
η	Efficiency
η_{lin}	Efficiency of an ideal linear voltage regulator
η_{sw}	Efficiency of a switched mode voltage regulator
η_{DC-DC}	Efficiency of a DC–DC converter
ω	Angular frequency
ψ	Charging fraction of a capacitive converter topology
κ_T	Output capacitance topology factor
γ	Ratio between ideal and actual VCR of a capacitive DC-DC converter
β	Output impedance balancing factor
ε	Electric permittivity
Δ_{vSSL}	Ripple in SSL
Δ_{vFSL}	Ripple in FSL
Δr_i	Ripple on a capacitor with index i

Chapter 1
Introduction

Our society is governed by computers and other electronic appliances. The technology revolution of the past century has led to an unseen living comfort and everyday life would not be the same without the electronic appliances that entered our life during the past decades.

In Fig. 1.1 some appliances in our surroundings are depicted. From a power-electronics engineer's point of view, these appliances are roughly divided in the following categories: mains-powered, battery-powered and Radio-Frequency(RF)-powered.

Mains-powered appliances are the devices that consume large amounts of power ($>10W$), for example home appliances, but also data servers. Although the electronic circuits , inside these devices, require a DC voltage supply (12–3.3 V) they are connected with the AC mains (230 V). Obviously an interface is required that turns the AC voltage into a DC voltage.

Battery-powered devices are those that require a certain degree of user mobility: laptops, cell phones, but also pacemakers. The most important specification of these devices is their autonomy: the lower the power consumption, the higher their autonomy considering an equal amount of energy available in the power source. In many cases Li-ion batteries, with a nominal output voltage of 3.6 V, are used as a power source. But part of the electronics inside the devices runs at 1.8 V, while other parts of the system require a 5 V supply. The application must comprise a circuit that converts the battery voltage into both the supply voltages.

The RF-powered devices are appreciated because of the perspective of battery-less operation. For example, Radio-Frequency Identification-tags are used in warehouses to identify goods without the need to physically examine them. They are scanned periodically by means of an RF signal. The latter is converted into a stable DC supply. But since power is transferred through the air when needed, no battery is required if the energy is stored on a capacitor for example. Because of this the volume is reduced and the autonomy is at its maximum.

Previous paragraphs demonstrate that: in most electronic appliances there is a striking mismatch between the power source and the electronics inside the appliance.

T. Van Breussegem and M. Steyaert, *CMOS Integrated Capacitive DC–DC Converters*, Analog Circuits and Signal Processing, DOI: 10.1007/978-1-4614-4280-6_1,

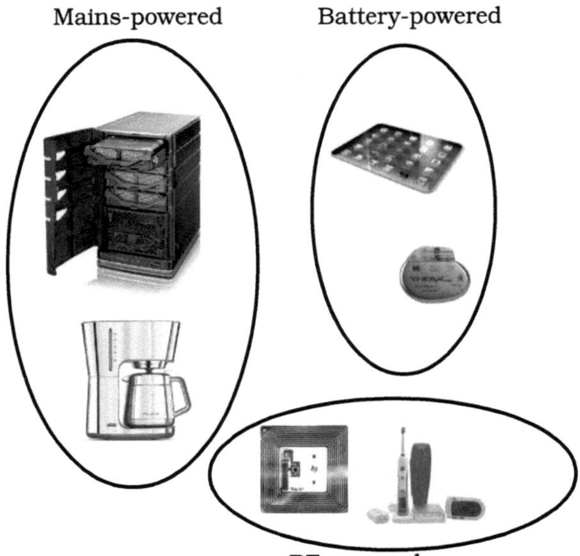

Fig. 1.1 Electronic appliances of all kinds surround us in everyday life. Three distinct categories are observed: mains-powered, battery-powered and RF-powered

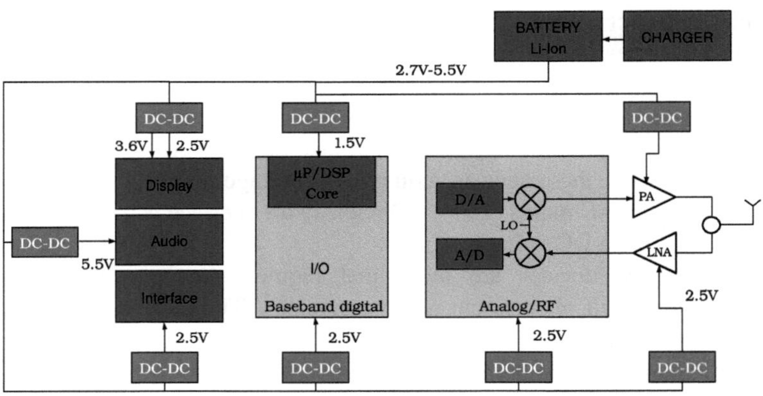

Fig. 1.2 Generic cell phone architecture

Power-electronics engineers device circuits that interface between the power source—whether it is the mains, the battery or other—and the electronic circuits inside.

As an example the power architecture of a cell phone is demonstrated in Fig. 1.2. The cell phone is a diverse assembly of sub-blocks, each sub-block dedicated to a distinct aspect of the transceiver functionality. The battery provides a variable voltage, this voltage depends on the energy amount stored in the battery, the output impedance of the battery and the load current drawn by the cell phone. Each of the sub-blocks of the architecture requires a specific operating voltage to achieve an optimal overall performance. In this architecture at least four different supply rails are required: 1.5, 2.5, 3.6 and 5.5 V. They have to be generated out of a single battery.

The DC–DC converters are building blocks that convert the battery voltage into voltages that comply with the building block specifications. Moreover the DC–DC converters also isolate the supply rails from each other to minimize interference between the blocks.

The DC–DC converters not only match different building blocks with different supply voltages, at the same time they provide more freedom for the circuit designers. By providing an interface between the power source and the circuits, the design of the latter is made less dependent on the power-source characteristics and this enables optimization of the circuit towards higher performance, low power consumption or both. And this is the primary source of concern in modern electronic appliances: how to reduce power consumption of the electronic circuits while maintaining equal performance.

This first chapter provides a broad introduction that outlines the context of this work. In Sect. 1.1 the System-on-Chip (SoC) concept and the use of power management in a SoC is motivated. Section 1.2 presents a number of power-management techniques. Eventually DC–DC conversion and the requirements for DC–DC converters used in a power-management context are covered in Sect. 1.3. Moreover, the different types of DC–DC converters are presented in this section. Each of these techniques is analyzed and compared to each other. Section 1.4 gives an overview of the trend-setting DC–DC converters found in recent scientific literature. Their performance and characteristics are discussed and compared to each other. In Sect. 1.5 an outline of the work is provided. And conclusions are drawn in the final section.

1.1 System-on-Chip Power Management

Only four decades ago, complex electronic systems were entirely built from multiple discrete-type components and mounted on a Printed-Circuit Board (PCB). This approach led to bulky electronic devices and was a serious obstacle when trying to reduce the volume of electronic appliances. Under the influence of the development of solid-state circuits a new concept, called a SoC, grew in importance. Instead of building a system by means of interconnecting different chips or components, as much functionality as possible— previously dispersed over different chips—was integrated on the same chip. By choosing a SoC, both size and eventual cost of these systems was drastically reduced. This led to a new paradigm in electronics, *The Integration Paradigm*:

> The Integration Paradigm looks for opportunities to integrate more and more functionality on a single silicon die or in the same integrated circuit (Steyeart and Vancorenland 2002). Nowadays the paradigm especially proves it strength in RF circuits, with mobile communication as its major driver.

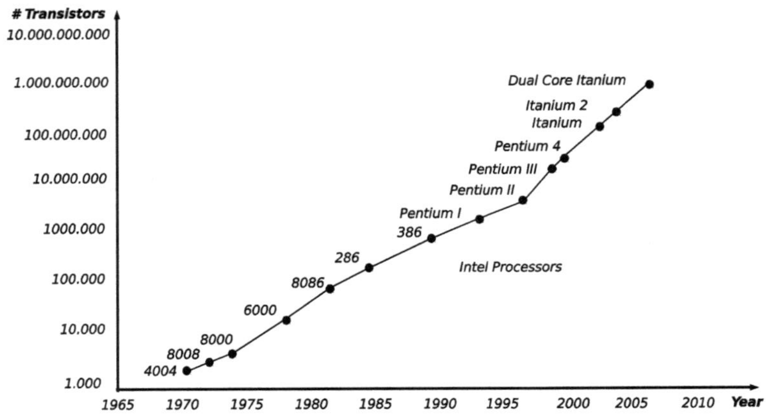

Fig. 1.3 Moore's law demonstrates the increase of transistor density over the past decades. The law states that transistor density doubles each 18 months

Complementary-Metal-Oxide-Semiconductor (CMOS) technology demonstrated to be an excellent technology to build integrated circuits and thanks to the ever-decreasing feature size of the state-of-the-art CMOS technologies (The international technology road map 2009), larger circuits can be constructed on the same chip area. In Fig. 1.3, Moore's law is demonstrating the increase of the transistor-count—transistors are the building blocks of advanced circuits in CMOS—in consecutive processor-designs. In spite of the continuous advances in circuit design and semiconductor physics, the operation of modern SoC's is threatened by a number of issues related to power. The following paragraphs elucidate the three main issues faced by system/power architects: the proliferation of power density, the voltage gap and the energy gap.

1.1.1 Power Density

A microprocessor is a schoolbook example of a full-grown SoC. It consists of dozens of sub-blocks each with their specific function and often each of the blocks requires a separate supply voltage. The tremendous increase of transistor density in computer chips has led to large increase in power dissipation of the circuits. This is illustrated in Fig. 1.4 by the power density—this is the power dissipation normalized with respect to the area where this power is dissipated—of succeeding generations of computer chips. Unless measures are taken, the power density of future processors runs out of hand. The heat dissipation associated with the power, can not be supported by the technology neither by the power supply used in these computers. Moreover the heat generated decreases the performance of the processor. Hence intensive cooling strategies are introduced to anticipate on this: heat sinks, water cooling, a heat pipe

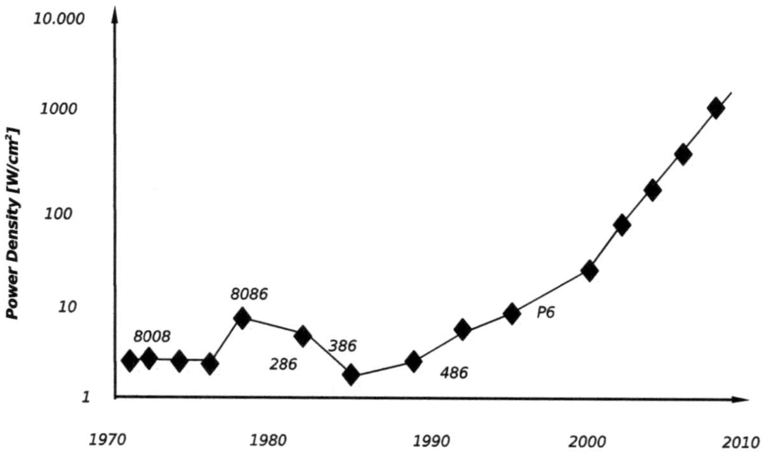

Fig. 1.4 The evolution of power dissipation over the past generations of processors

Table 1.1 Battery types and nominal voltages

Type	$V_{nominal}$
NiMH	1.2 V
Li-Ion	3.6 V
Ni-Cd	1.2 V

and even liquid immersion cooling (Ellsworth 2004; Chu et al. 2004). These techniques not only introduce a large hardware cost but also additional power loss. This concern is demonstrated by the following quote by the CEO of Google in 2002:

> *What matters most to the computer designers at Google is not speed, but power—low power, because data centers can consume as much energy as an entire city.*—Eric Schmidt, CEO of Google

1.1.2 Voltage Gap

Besides the power density issues, another phenomenon called the voltage gap is rising significant problems. The voltage gap concerns the mismatch between the nominal voltage of the commonly used batteries and the maximum supply voltages of the state-of-the-art electronic circuits. In Table 1.1 a short overview is given of the most commonly used battery-types and their nominal output voltage. Nominal battery voltages range from 1.2 to 3.6 V. But the maximum and minimum voltages of a Li-Ion battery—for example—are as wide as 2.5 to 4.5 V. Most of the electronic circuits not only require a lower voltage but also a more stable voltage.

Table 1.2 Maximum voltage of the native devices in state-of-the-art CMOS technologies

Technology	V_{max}
CMOS 0.25 μm	2.5 V
CMOS 0.18 μm	1.8 V
CMOS 0.13 μm	1.2 V
CMOS 90 nm	1.2 V
CMOS 45 nm	1 V
CMOS 32 nm	1 V

In Table 1.2 the maximum voltages of the state-of-the-art CMOS technologies, used for integration of electronic circuits, are given. These voltages range from 2.5 to 1 V in the most recent technologies. Where the battery voltages are dictated by the battery chemistry,[1] the break-down voltages of the state-of-the-art CMOS technologies are governed by the laws of physics.[2] In order to match these differences, an interface circuit is required to convert the battery voltage into the appropriate voltage to supply the electronic circuits.

1.1.3 Energy Gap

Besides the obvious technical constraints—chemistry versus physics—issues related to human behavior play an at least equally important role. People tend to expect more and more from their electronic appliances. This is demonstrated in Fig. 1.5 based on data from Texas Instruments.[3] This figure compares the evolution of available energy in a 10 cc battery pack with the energy required to meet the consumer's expectations. Since 2007, the energy a battery of reasonable size can contain, does not meet the energy required to fulfill the consumers expectations. These expectations include mobile connectivity and video transfer at high speed (Delagi 2010). Although continuous research effort is directed towards increasing the energy density of batteries, an at least equally probable approach is to reduce the power consumption of the circuits. To achieve this a number of techniques has been developed to cut the power loss in large applications by applying power management by means of the so-called power-management techniques.

[1] The battery voltages depend on the electrochemical properties of materials.

[2] The maximum voltage rating of CMOS processes is determined by the maximum electrical field (expressed in V/m) the gate oxide can withstand.

[3] TI is one of the most prominent IC manufacturers in the world.

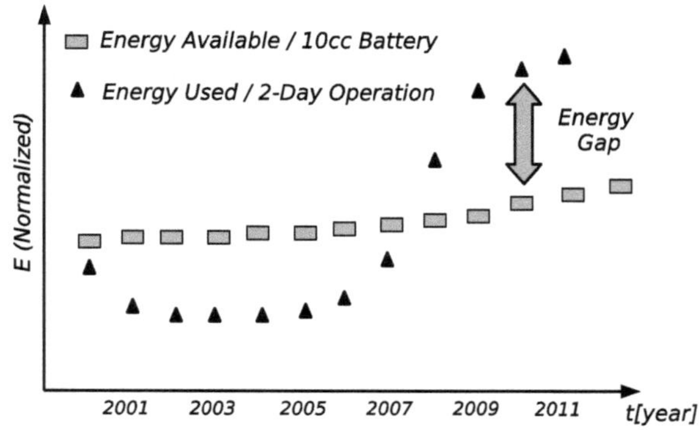

Fig. 1.5 Since 2007, the energy a battery of reasonable size can contain does not meet the energy required to fulfill the consumers expectations (Delagi 2010)

1.2 Power-Management Techniques

The three issues, presented in the previous section, demonstrate that appropriate techniques are required for SoC's to comply with the ever increasing power needs. This can be approached from a device point of view, by developing devices less prone to power loss, or by engineering circuits that achieve equal performance while using less power. First we discuss in short the main sources of power loss in electronic circuits and next a selection of power saving techniques is presented.

1.2.1 Power Consumption in CMOS

Power consumption in CMOS has multiple origins. First one can differentiate between the power consumed in analog circuits and the power consumed in digital circuits. In most analog circuits the power consumption is dominated by the bias currents $I_{bias,i}$ of the transistors. Even if the analog circuits are not actively used, these bias currents continuously dissipate power:

$$P_{Analog} = \Sigma_i I_{bias,i} V_{supply} \tag{1.1}$$

This opposed to digital circuits in CMOS that ideally only consume power when activated. The power loss in digital circuits is proportional with the square of the supply voltage V_{supply}, the amount of (gate-)capacitance in the circuit $C_{gate,i}$, the clock frequency f_{clk} and the degree of activity of every gate $\alpha_{activity,i}$:

$$P_{Digital} = \Sigma_i \alpha_{activity,i} C_{gate,i} V_{supply}^2 f_{clk} \tag{1.2}$$

This gives digital circuits a clear advantage over analog circuits. Moreover, the power loss in digital circuits is proportional to the square of the supply voltage. This

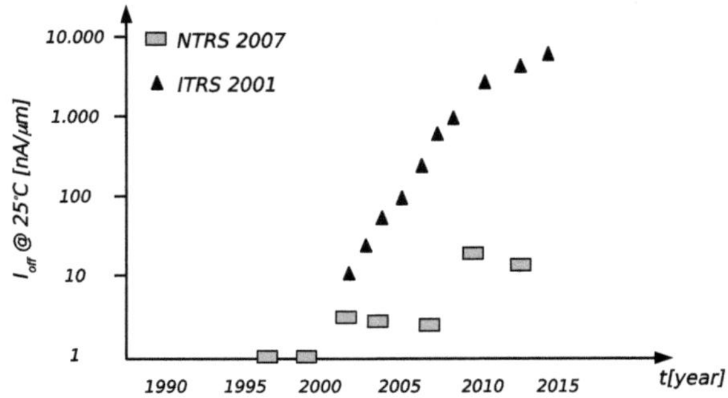

Fig. 1.6 Leakage Current Trend comparing prediction from 2001 with observation/prediction in 2007

makes it very attractive for designers to aim for lower supply voltages. Especially since digital circuits have a large signal-to-noise margin by nature.

Additionally to the biasing power loss in analog circuits and the gate switching loss in digital circuits, a number of secondary effects increase the power loss even more. In the following paragraphs two of them are discussed: sub-threshold leakage and shoot-through-current loss.

Sub-Threshold Leakage

In early models, the transistor is supposed to be off when the gate-source voltage V_{GS} is smaller than the threshold voltage V_{th}. Actually even in this so-called cut-off region, a significant current $I_{Sub-V_{th}}$ flows through the transistor:

$$I_{Sub-V_{th}} = \frac{W}{L_{eff}} I_s e^{[\frac{1}{\eta V_{th}}(V_{gs}-V_{tho}-\gamma_{tech} V_{sb}+\beta V_{ds})]}(1-e^{\frac{V_{ds}}{V_{th}}}) \qquad (1.3)$$

From a power-consumption point of view, sub-threshold conduction in digital circuits is a major source of power loss. The sub-threshold leakage current introduces a static power loss that can not be neglected. Fig. 1.6 demonstrates the increasing importance of sub-threshold conduction over the past decades. This graph shows the amount of leakage current for the most recent technology nodes. Although the prediction from ITRS2001 provided a more pessimistic outcome than the actual evolution monitored by NTRS2007.

Shoot-Through Current Loss

The shoot-through current loss denotes the power loss due to the current that flows in a CMOS inverter (or other complementary logic gate) from the positive supply to the ground during a change at the input (or one of the inputs). In Fig. 1.7 the transient from a low to high input of an inverter is shown. In the first phase the

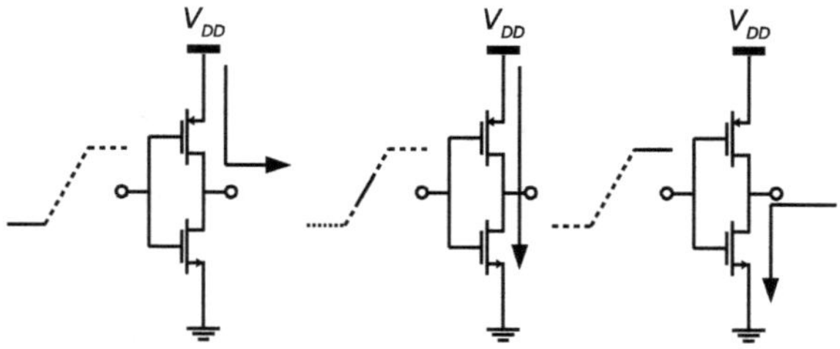

Fig. 1.7 The shoot-through current loss denotes the current that flows in a CMOS inverter from the positive supply to the ground during a transient at the input

PMOS device is conducting and the NMOS is off. In the final phase the PMOS is off and the NMOS is on. During the change from low to high, both devices are on, since for both devices the voltage between the gate- and source terminal is higher than the threshold voltage V_{th}. Although this phenomenon occurs during a very short time period, considerable amount of energy is dissipated in the MOS devices and is not directed towards the load of the inverter.

1.2.2 Clock Gating

The most straightforward technique to reduce power consumption in a digital system, is by shutting down the clock: this technique is referred to as clock gating. In most synchronous systems, a large part of the flip flops are clocked disregarding they are processing data or not. Therefore it is a viable approach to shut down the clock in an idle part of the chip. Since the supply voltage remains high, data retention is ensured. This technique requires insight in the system's operation, but is a quite simple approach. In Fig. 1.8 a schematic representation is given how clock gating can be implemented by adding a number of gates in the clock tree. Obviously a hardware overhead is created.

Power management in a 10-core Westmere Intel processor. The Westmere (Sawant et al. 2011) is a processor for the enterprise market aiming for high-performance computing. It comprises 10 distinct processor cores. Power management is one of the key features of this product released in 2011. The power management not only relies on per-core power gating but also on the power reduction in the critical periphery (the so-called *uncore*). The latter is achieved by shutting down bias currents to on-chip DLL's, transmitters and phase-interpolators.

Clock gating can be performed on different levels: coarse-grained or fine-grained. Coarse-grained means on the level of modules and is often software-controlled, while fine-grained is on the register-level and thus it is a more hardware-oriented technique. The choice between both approaches is influenced by the trade-off between the

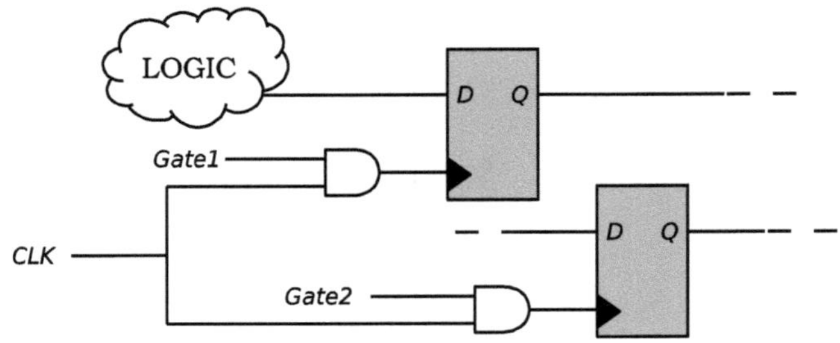

Fig. 1.8 Clock gating on register level, this approach introduces high hardware overhead but also large energy savings

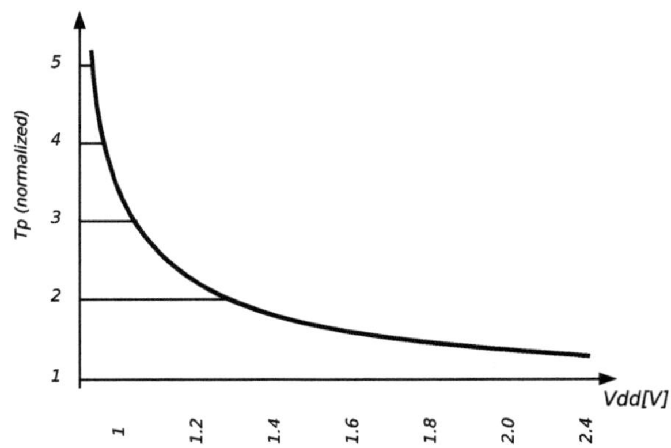

Fig. 1.9 Propagation delay of a CMOS inverter in 0.25 μm technology as a function of the supply voltage (Rabaey et al. 1996)

hardware overhead and the potential power-loss reduction. For module-level gating the overhead is small, the potential power saving small, while for register-level clock gating the overhead is large and the potential power saving as well.

1.2.3 Voltage and Frequency Scaling

Digital circuits and especially microcontrollers and microprocessors are used in a wide variety of applications. In many cases the maximum clock frequency of such a microprocessor is not necessary to meet the user requirements. In terms of microprocessors it is very beneficial to adapt the clock frequency to the processor's computational load. When the load is low the frequency is decreased to a sleep mode for example. During calculation tasks, the clock frequency increases and the processor runs in a high-performance mode. Due to the linear relationship between power consumption and clock frequency in Eq. 1.2, a large benefit can be earned by addressing this technique.

Equation 1.2 reveals an even stronger relationship between switching power loss and the supply voltage: the power loss is proportional with the square of the supply voltage. Instead of scaling clock frequency, supply voltage can be scaled or the combination of frequency and supply voltage. In Fig. 1.9 it is demonstrated how the propagation delay of a CMOS inverter [4] scales with the supply voltage: for small supply voltages the propagation delay increases exponentially. Hence if the supply voltage is scaled down, the circuit's maximum clock frequency must be scaled along with the supply voltage. Otherwise set-up or hold-time violations occur in the circuit due to the excessive propagation delay with respect to the frequencies' period. Voltage scaling is thus inevitably combined with frequency scaling.

DVS was first introduced in the StrongARM/XSCALE processor (Clark ISSCC 2001). It is a 16.77mm^2 processor built in 0.18 μm CMOS technology. The performance is characterized at 1.65 V/800 MHz, 1.3 V/600 MHz and 0.75 V/200 MHz. The power consumption in these modes is ranging from 900 mW down to 50 mW. This demonstrates the potential reduction in power consumption thanks to the DVS technique (Clark et al. 2001).

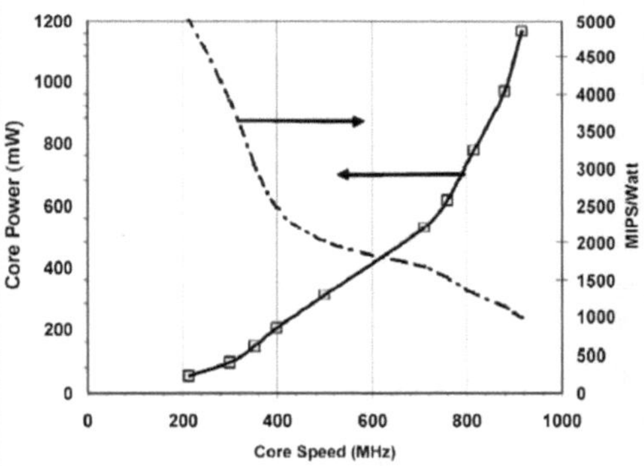

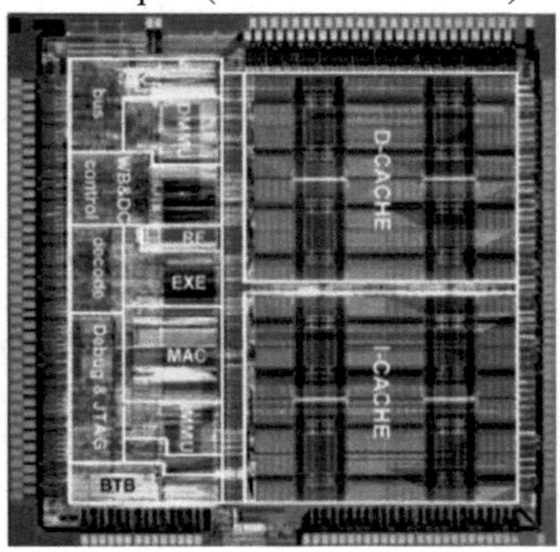

Dynamic Voltage Scaling

Dynamic Voltage Scaling (DVS) (Von Kaenel et al. 1990) is a power-management technique that uses voltage scaling as a tool to reduce the power consumption in digital circuits. It is based on the observation that a digital block (for example a microprocessor) has a wide varying workload and that scaling the clock frequency together with the supply voltage has not necessarily an influence on the *perceived* performance of the processor while the power consumption is reduced significantly.

DVS, schematically represented in Fig. 1.10, is controlled by a PMU. This PMU gets workload information from the processor. Based on a Look-Up Table (LUT) a predefined frequency-voltage combination is programmed by the PMU (Brodersen et al. 2000). This technique requires a conservative approach since it is a Feed-

[4] The propagation delay is the amount of time between the input of a logic gate is stable and the instant that the output of that logic gate is stable.

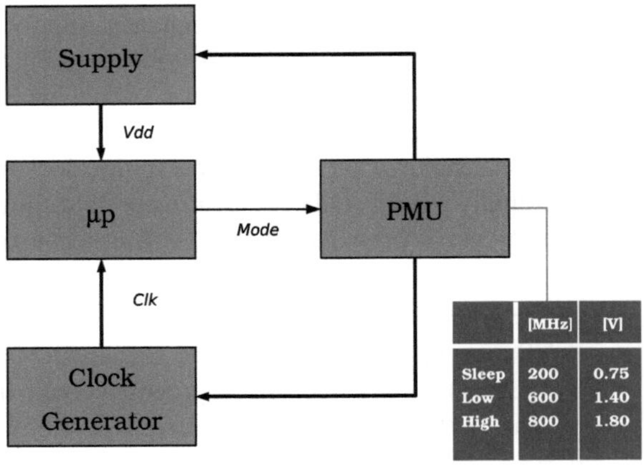

Fig. 1.10 Schematic representation of a DVS power-management system. The processor sends workload information to the PMU. This PMU controls both the processor's clock frequency and the supply voltage. The frequency and supply voltage is determined based on a hard-coded LUT

Forward (F^2) technique. The predetermined voltage/frequency combinations should meet the throughput specifications in a broad range of temperature variations, if subject to process variations and even under the influence of aging.[5]

Adaptive Voltage Scaling

It is clear that the Feed-Forward nature of DVS leads to a conservative implementation of the voltage-scaling approach. The voltage-scaling approach can be optimized by using a feedback realization instead: the technique is then called Adaptive Voltage Scaling (AVS) (Brodersen et al. 2000). The difference is that AVS monitors the processors performance—by means of a replica circuit—and adjusts both voltage and frequency accordingly. By doing so the effect of temperature and other variations are taken into account and design/operating margins are much smaller. This results in even less power loss. Figure 1.11 represents the setup of a AVS controlled microprocessor. In many cases (Brodersen et al. 2000), the delay in an oscillator is used to define the circuit's performance given the system's environment. The replica should be closely matched to the processor—both in nature as spatially—to mimic the behavior of the processor.

1.2.4 Adaptive Voltage Body Biasing

In circuit design the gate-, drain- and source-terminals of a transistor are regarded to as the most important ones, although manipulating the bulk terminal has significant influence on the transistors performance. The bulk voltage with respect to the source

[5] Transistor parameters are subject to aging effects. This means that the transistor parameters change over time, which may results in performance degradation.

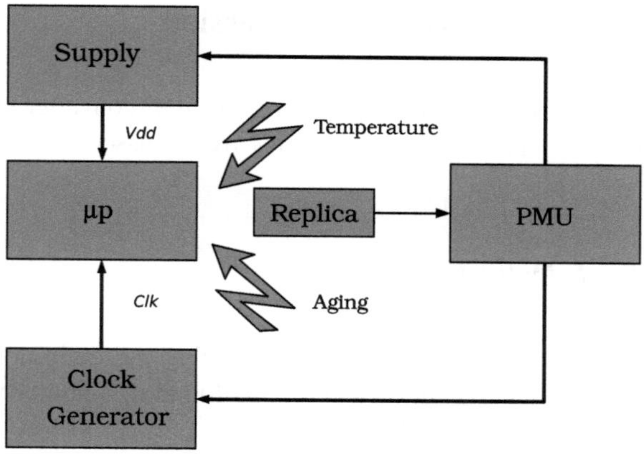

Fig. 1.11 Adaptive Voltage Scaling is a feedback technique that monitors the processor's behavior and controls the clock frequency and supply voltage accordingly

voltage influences the threshold voltage V_{th} of the transistor that is in turn one of the critical parameters for both analog and digital circuit design. By manipulating the bulk voltage the transistor characteristics are altered.

When focusing on the digital circuits, both static and dynamic body-biasing is addressed to improve circuit performance. Static body-biasing (SBB) involves biasing with various fixed voltages in a predetermined way. Again it is a Feed-Forward technique to improve the device performance. Adaptive body-biasing (ABB) monitors the performance and intervenes if necessary. As is demonstrated in Tschanz et al. (2002), forward ABB increases the performance of weak silicon by reducing V_{th}. Reverse ABB is able to decrease the leakage power of strong silicon by increasing V_{th}. By using this approach, it is demonstrated in Tschanz et al. (2002) that the yield in processor fabrication can be improved. By using ABB the processor manipulates the erroneous devices and by doing this finally meets the performance requirements.

1.2.5 Analysis

The previous paragraphs demonstrate a number of frequently used power-management techniques. They range from static predefined circuit manipulations (body biasing) over real-time programmed (DVS) to closed-loop self-controlled techniques (ABB and AVS). All these techniques have a single common feature and that is: the manipulation of the supply voltage of the circuit.

Manipulation of the supply voltage is achieved by inserting a Voltage Regulator Module (VRM) between the regular supply and the loading circuit. Actually this VRM is nothing more but a well controlled DC–DC converter. The most obvious incarnation (considering a conversion from a high voltage to a lower one) is the linear regulator but a set of alternatives exists under the form of switched-mode power supplies. The next section presents an overview of the DC–DC converters which can be used in a VRM.

1.3 DC–DC Voltage Conversion

First the qualification of the term DC–DC converter is discussed, next a number of DC–DC converter characteristics are discussed.

1.3.1 Definition

There is a lot of ambiguity concerning the definition of a DC–DC converter. In the box underneath the definitions found in a number of standard-works on power electronics, are cited.

Power Electronics by Mohan et al. (2002), Undeland and Robbins: A DC–DC converter is a single power-conversion stage that may perform DC-to-DC voltage conversion. It utilizes semiconductor devices controlled by signal electronics and possibly energy-storage elements such as inductors and capacitors.

Fundamentals of Power Electronics by Erickson and Maksimovic (2001)

In general a switching converter contains power input, control input ports and a power output port. ... One of the several basic functions can be performed. In a DC–DC converter the DC input voltage is converted to a DC output voltage having a larger or smaller magnitude, possibly with an opposite polarity or with isolation from the input and output references.

Principles of Power Electronics by Verghese et al. (1991) ... a circuit that turns a DC input into a DC output..

Elements of Power Electronics by Philip T Krein (1998) ... that uses lossless elements only..

It is clear that these authors stress different aspects of the conversion process. In (Mohan et al. 2002) the controllability is an important item. In (Philip T Krein 1998) the emphasis is put on the fact that ideally the components are lossless, while this is not necessarily required according to the definition of Mohan et al. (2002) and Maksimovic and Erickson (2001) or Varghese et al. (1991). The only consensus found between the definitions from these authors is the obvious requirement that a DC–DC converter turns a DC voltage at the input into a DC voltage at his output. For sake of clarity the following definition is used in this work:

Definition:
A DC–DC converter is a circuit that has as primary feature to convert a DC voltage at the input terminals into another DC voltage at the output terminal.

This has as a consequence that both linear and switched-mode (inductive and capacitive-type) DC–DC converters are considered. But first the main requirements of a DC–DC converter are analyzed.

1.3.2 Requirements and Characteristics

Previous sections demonstrate that a DC–DC converter is used for a broad range of tasks and applications. The most critical role of the DC–DC converter is to match the supply voltage with the circuit voltage and to relax the relationship between these voltages. To fulfill this task a number of characteristics are listed which are used to describe the performance of a DC–DC converter but also to compare converter prototypes with each other. Since different converters serve different needs, not every converter excels on each characteristic. For some converters part of these characteristics are ignored while others are regarded to as their primary specification.

The DC–DC converter characteristics are divided in two main categories: the static and the dynamic characteristics. The static characteristics denote the nature of the converter in steady-state regime, while all environmental conditions are fixed and the transient behavior is settled out. The dynamic characteristics denote the nature of the converter during transients or under the influence of past transients.

Static Characteristics

The static requirements are strongly connected to the nature of the converter stage itself and not to the control technique used by the DC–DC converter. The most important characteristics are:

Voltage-Conversion Ratio The Voltage-Conversion Ratio (VCR) is the ratio between the output voltage V_{out} and the input voltage V_{in} of the conversion stage.

$$VCR = \frac{V_{out}}{V_{in}} \tag{1.4}$$

A DC–DC converter with a VCR larger than one is called an up converter. Examples are the Boost Converter (Philip T Krein 1998) and the Dickson Charge Pump (Dickson 1976). When the VCR is smaller than one, it is a down converter, for example the buck converter and a linear regulator. For a negative VCR, the converter is one of the inverting types. Most DC–DC converters provide a broad range of VCR's, but the most simple types are limited to either up or down conversion.

Noise Ideally a voltage source can be modeled as an ideal voltage source with a zero output impedance. A proper supply voltage provides a noise-free DC–DC voltage regardless of the load current. Most voltage sources have an intrinsic (sometimes

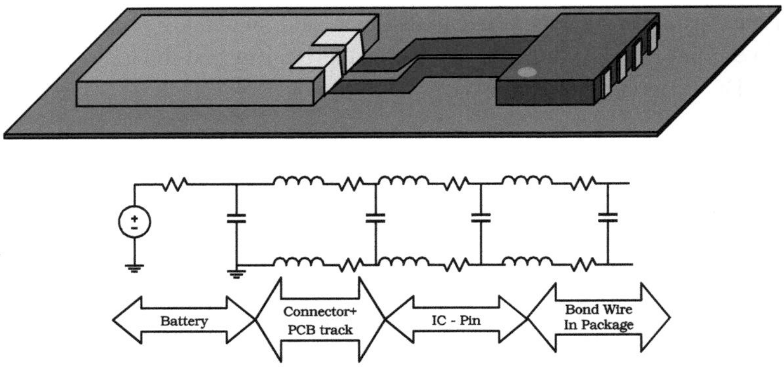

Fig. 1.12 Lumped Model of the physical connection between a battery and a chip mounted on a PCB

varying) output impedance.[6] Moreover every physical intervention in the connection between the voltage source and the load, adds a parasitic component to the final output impedance. In Fig. 1.12 the model of a realistic connection between a voltage source (battery) and a chip is demonstrated. First of all the battery is modeled as an ideal voltage source with a varying output impedance. The connector between the battery and the PCB track adds parasitic capacitance, resistance and inductance. So does the PCB track itself, the pin of the chip package and the bond-wire that connects the silicon die and the internal package lead.

If a varying current flows through the series and parallel-connected RLC-ladder, the DC voltage of the source is turned into a voltage that exists of a large DC component and an AC signal. This AC signal is called the input noise.

The same effect is observed at the output of the DC–DC converter: noise is invoked by the combination of a non-zero impedance and a pulsating current. If a large output capacitor is used, the AC-output impedance is reduced and the noise is decreased. In general, adding capacitors reduces the noise but at the cost of control bandwidth and increased bill of materials.

Efficiency The efficiency η of a DC–DC converter is the foremost important characteristic of a DC–DC converter. It is the ratio of the converter's output power P_{out} and the input power P_{in}. The difference between input and output power is the power loss: P_{loss}. The efficiency is calculated as follows:

$$\eta = \frac{P_{out}}{P_{in}} \tag{1.5}$$

This equation is also equal to:

$$\eta = \frac{P_{out}}{P_{loss} + P_{out}} \tag{1.6}$$

[6] Batteries have an output impedance that depends on the energy contained in the battery.

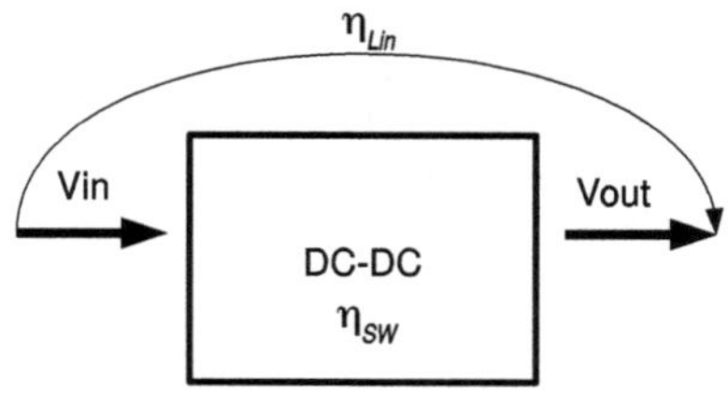

Fig. 1.13 A single-stage DC–DC converter by means of a switched-mode DC–DC converter

In an ideal case, there is no power loss and the efficiency reaches 100 % under all circumstances. If real components are used in a DC–DC converter, these components introduce resistive losses and electrical power is partially transferred into heat. Many other sources of power loss are found in an actual DC–DC converter, these are analyzed in the forthcoming chapters.

Power Density is the output power of a DC–DC converter normalized with respect to the area required to perform the conversion:

$$PD = \frac{P_{out}}{A} \tag{1.7}$$

In fact power density is only rightfully calculated for planar implementations of DC–DC converters: monolithically integrated. For converters using discrete-type of converters, the volume should be taken into account.

Efficiency Enhancement Factor Until recently DC–DC converters were primarily qualified by means of their power density or their efficiency ratings. But especially for down converters efficiency ratings turned out to be an inadequate figure to qualify these converters. First of all, because there is no obvious relationship between the efficiency of the converter and the reduction in power loss due to the converter. Secondly because the efficiency should be normalized with respect to the conversion factor—this is demonstrated in the following paragraphs. First one has to recognize that a down-conversion DC–DC converter can be benchmarked with respect to the most straightforward down-converter: the linear regulator. The series linear regulator is a closed-loop circuit that dissipates the excess voltage in a pass device. This circuit has the drawback of executing this conversion at an efficiency that corresponds to the ratio between output and input voltage. As a consequence the theoretical maximum efficiency for linear regulators is small for conversions with a small VCR. In general, linear regulators can be build in a very compact way and operate at an efficiency very close to their theoretical maximum efficiency. Hence linear regulators are excellent vehicles to benchmark down-conversion DC–DC converters.

Each down-conversion DC–DC converter should perform better than a linear regulator. In Fig. 1.13 a single-stage DC–DC conversion by means of a switched-mode DC–DC converter is shown: η_{SW} represents the efficiency of a DC–DC converter and η_{lin} is the maximum efficiency of a linear regulator executing the same conversion. The Efficiency Enhancement Factor (EEF) is defined as:

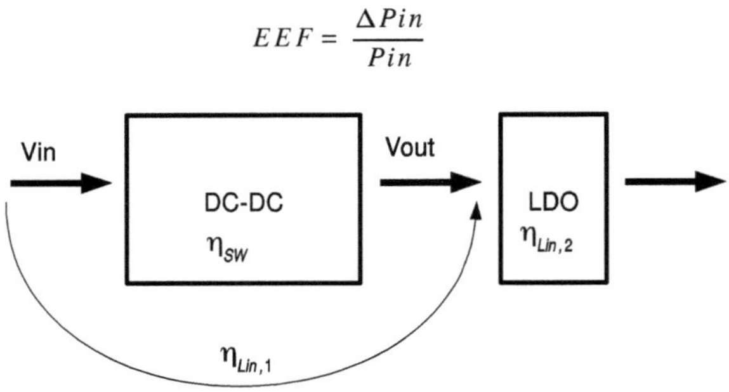

Fig. 1.14 A two stage DC–DC converter by means of a switched-mode DC–DC converter and a series linear regulator

$$EEF = 1 - \frac{\eta_{lin}}{\eta_{SW}} \tag{1.8}$$

It can be proven (Mike Wens and Michiel Steyaert 2011) that the EEF corresponds to the power loss reduction (ΔPin) due to the DC–DC converter normalized with respect to the power consumed by a hypothetical linear regulator (P_{in}) performing the same conversion as the DC–DC converter. Thus the EEF is a true measure of the power-loss reduction:

$$EEF = \frac{\Delta Pin}{Pin} \tag{1.9}$$

If the ripple at the output of the DC–DC converter (with $\eta_{sw} = \eta_{DC--DC}$) is too high, a series linear regulator is added. A two-stage DC–DC conversion is obtained, this is shown in Fig. 1.14, where the switched-mode converter deals with the large DC-voltage span and the linear regulator filters out the ripple. But this does not alter the EEF of the system. This is demonstrated by the following interpretation of the EEF in a cascaded system. The EEF of a standalone DC–DC converter is defined as:

$$EEF = 1 - \frac{\eta_{lin}}{\eta_{SW}} = 1 - \frac{\eta_{lin,1}}{\eta_{DC--DC}} \tag{1.10}$$

In this formula $\eta_{lin,1}$ is the maximum theoretical efficiency of an ideal series regulator that executes the same conversion as the DC–DC converter. If the DC–DC converter is followed by a series regulator with efficiency $\eta_{lin,2}$ the formulation is altered accordingly. But now $\eta_{lin} = \eta_{lin,1} * \eta_{lin,2}$ and also the converter efficiency is now $\eta_{DC--DC} * \eta_{lin,2}$: This reduces for the new system to:

$$EEF = 1 - \frac{\eta_{lin}}{\eta_{SW}} = 1 - \frac{\eta_{lin,1} \times \eta_{lin,2}}{\eta_{DC--DC} \times \eta_{lin,2}} \tag{1.11}$$

$$EEF = 1 - \frac{\eta_{lin,1}}{\eta_{DC--DC}} \tag{1.12}$$

This demonstrates that if a converter is followed by a linear regulator, the EEF of the final system (DC–DC + post regulator) equals the EEF of the standalone converter. Considering the operation point of the converter remains the same.

Accuracy It is desirable that the DC–DC converter is externally controllable and tracks the controlling signal as close as possible irrespective of the circumstances. This accuracy can be expressed in a number of ways, the most straightforward being:

$$e_{control} = \frac{V_{out}}{\kappa_{control} V_{control}} \times 100\,\% \tag{1.13}$$

$e_{control}$ is then the percentage error of the output voltage with respect to the control voltage. Taken into account that $\kappa_{control}$ is a scaling factor depending on the output-voltage-sensing method.

Dynamic Characteristics

The dynamic characteristics are influenced both by the converter stage as by the control method/ circuit of the DC–DC converter. The most fundamental characteristics are shown in Fig. 1.15.

Line Regulation Line regulation denotes the potential of a DC–DC converter to deal with variations in line voltage ($V_{in2} - V_{in1}$). By measuring the variation of the output voltage ($V_{out,in2} - V_{out,in1}$) at two distinct line/input voltages and normalizing this variation with respect to the line variation, a percentage line regulation is calculated:

$$R_{line} = \frac{V_{out,in2} - V_{out,in1}}{V_{in2} - V_{in1}} \times 100\,\% \tag{1.14}$$

Line Regulation is not necessarily linear over the whole input range, therefore the output variation is calculated for the maximum and minimum input voltage.

Load Regulation Load regulation expresses the potential of a DC–DC converter to deal with variations in load current. By measuring the variation in output voltage at two distinct load currents $I_{1/2}$ and normalizing this variation with respect to the load variation, the load regulation is calculated:

$$R_{load} = \frac{V_{out,I2} - V_{out,I1}}{I_2 - I_1}\,\Omega \tag{1.15}$$

Load regulation is not necessarily linear over the whole load range. Therefore the output variation is calculated for the maximum and minimum load current.

Bandwidth The bandwidth of a DC–DC converter describes how good the converter can deal with changes in load, line and control-signals. The load-regulation bandwidth gives the highest frequency of load variation that is tolerated without

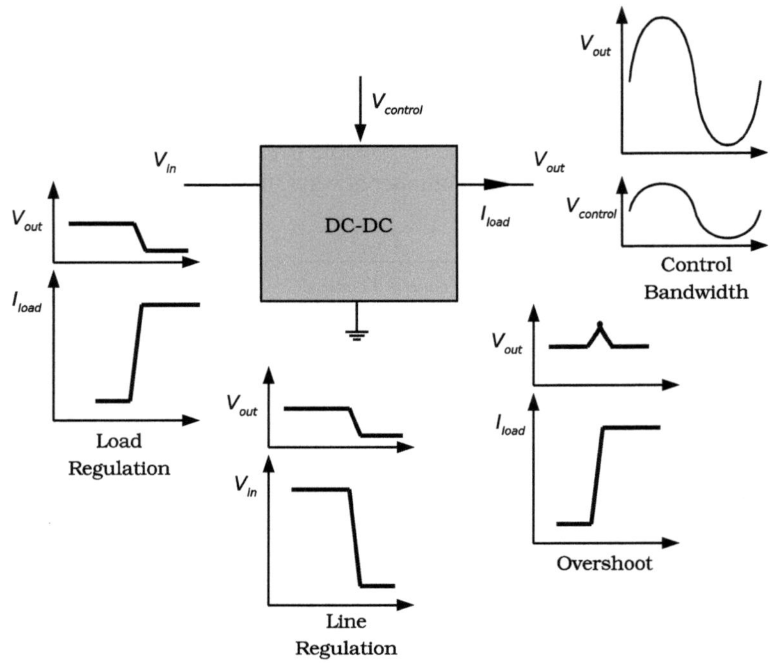

Fig. 1.15 Overview of a number converter specifications: load regulation, line regulation, overshoot and bandwidth

violating the other requirements. This feature is tested by applying a load step from minimum load to full load with predefined rise/fall- time.

For Point-of-Load (PoL) converters this is the utmost dynamic specification. The line-regulation bandwidth—which maximum frequency of variation is tolerated at the input of the converter—is of high importance for converters with multiple inputs.

Overshoot Overshoot and undershoot is the deviation from the nominal output voltage due to a transient in load-line or control. It should be exactly specified under which circumstances and operating points the overshoot occurs.

In the following sections the primary conversion techniques and their main characteristics are explored, detailed discussion follows in the next chapters.

1.3.3 Linear Series Conversion

Linear series conversion is a conversion technique that is restricted to down conversion. It includes conversion by means of a series pass device. This is demonstrated in Fig. 1.16. In the ideal case, the input current I_{in} equals the output current I_{out} and the efficiency η_{LR} of the linear regulator is expressed as:

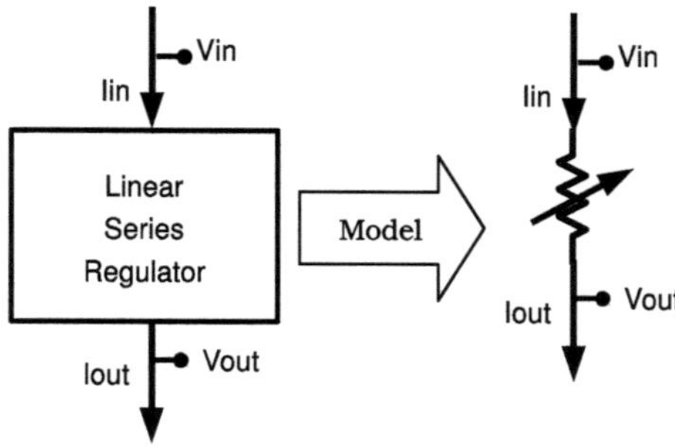

Fig. 1.16 Schematic representation of a series linear regulator and its model

$$\eta_{LR} = \frac{P_{out}}{P_{in}} = \frac{V_{out} I_{out}}{V_{in} I_{in}} = \frac{V_{out}}{V_{in}} \tag{1.16}$$

In an actual implementation achieving the regulation requires a control loop. This control loop introduces an additional power loss term P_{loss}:

$$\eta_{LR} = \frac{P_{out}}{P_{in} + P_{loss}} < \frac{V_{out}}{V_{in}} \tag{1.17}$$

The model of a linear series regulator is an adaptive resistor. The resistance value of this resistor must be controlled in real time to ensure a constant output voltage even if the output current is changing. Given an excellent control loop, the linear regulator is the preferred conversion solution. Mainly since the conversion is achieved using a minimum of chip area and a minimum number of passive components: the simplest incarnation only requires a single buffer capacitor.

The main drawback of a linear series regulator follows from Eq. 1.16. According to the definition of the linear regulator's maximum efficiency, this efficiency is equal to the VCR. For small VCR's the maximum achievable efficiency is also small. Moreover no up conversion can be achieved by means of a linear regulator. To overcome these drawbacks, a switched-mode converter is used. Linear regulators are discussed more in detail in Chap. 4.

1.3.4 Capacitive Conversion

DC–DC converters using nothing but switches and capacitors to perform the voltage conversion are known as capacitive DC–DC converters.

Capacitors

Capacitive conversion relies on the energy-storage capabilities of capacitors. Capacitors are passive electric components that store energy in an electric field ($\vec{E}$) under the form of charge on two conductors separated by an insulator. The capability to

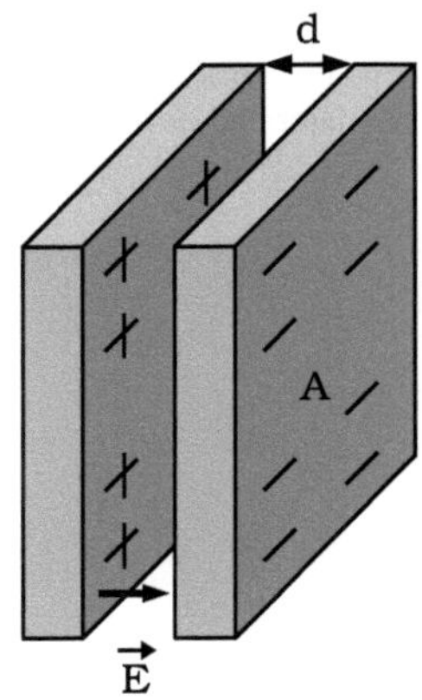

Fig. 1.17 Schematic representation of a parallel plate capacitor

store charge is expressed as the capacitance (C) of the capacitor. When a potential difference V is applied across the plates, a charge Q_+ is placed on one plate and Q_- on the other. By definition:

$$C = \frac{Q}{V} \tag{1.18}$$

The capacitance of a capacitor or either other combination of objects depends on their form and size. For a parallel-plate capacitor in Fig. 1.17, the capacitance depends on the permittivity ϵ of the insulator between the plates, the distance between the plates d and the area A of the plates:

$$C_{ParallelPlates} = \frac{\epsilon A}{d} \tag{1.19}$$

For a broad range of common capacitors and capacitance calculations of objects, the parallel plate approximation is used. For more complex structures, hand calculation is often omitted and dedicated tools (p.e. FastCap) are used.

When an electric field is built up between the capacitor's plates and this field can accelerate charge, a capacitor is capable of doing work and thus must contain energy. The work dW that is performed by moving an additional amount of charge dq from one plate to the other is:

$$dW = Vdq = \frac{q}{C}dq \tag{1.20}$$

The total work performed to increase the charge from zero to Q is then:

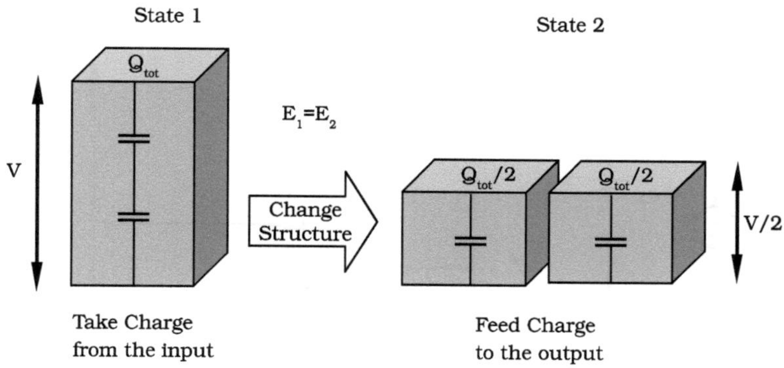

Fig. 1.18 The concept of capacitive conversion, from a charge point of view

$$W = \int_0^Q \frac{q}{C} dq \tag{1.21}$$

$$= \frac{Q^2}{2C} \tag{1.22}$$

$$= \frac{CV^2}{2} \tag{1.23}$$

Of course it is not enough to store energy on a capacitor, the capacitor must be able to transfer energy as well. And this as efficiently as possible.

Energy Transfer

A capacitive converter uses capacitors to transfer charge from the input to the output of the converter. By doing so the amount of charge is conserved but the means of energy storage is altered. This is demonstrated conceptually in Fig. 1.18 for a converter that divides the input voltage by a factor of two: both structures contain equal amount of energy. In the first state the energy is stored on a series connection of capacitors. In the second state the energy is stored on a parallel structure. But despite the conservation of energy, a change in voltage across the structure is observed. This shows that a structure of capacitors can achieve a voltage conversion. The conversion details are discussed in the forthcoming chapters.

In a real capacitive DC–DC converter energy is transferred from the input supply to the load by means of capacitors. The energy transfer from a supply to a single capacitor [7] is modeled as charging a capacitor by a voltage supply through a resistor in Fig. 1.19.

---

[7] For sake of simplicity charging a single capacitor is analyzed but the conclusions can be extended to transfer between two capacitors or more complex structures

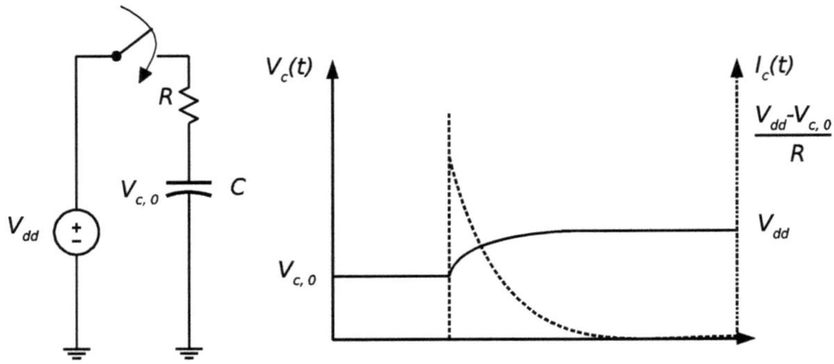

Fig. 1.19 Charging of a capacitor

The process of charging a capacitor is described by the following differential equation:

$$-V_{dd} + RC\frac{dV_C}{dt} + V_C = 0 \tag{1.24}$$

By solving towards $V_C(t)$ and deriving the current through the capacitor $I_C(t)$, the following expressions are given:

$$V_C(t) = V_{dd} - (V_{dd} - V_{C,0})e^{\frac{-t}{RC}} \tag{1.25}$$

$$I_C(t) = \frac{V_{dd} - V_{C,0}}{R}e^{\frac{-t}{RC}} \tag{1.26}$$

The power transferred to the capacitor is:

$$P_C(t) = V_C(t)I_C(t) \tag{1.27}$$

$$= \frac{V_{C,0}V_{dd} - V_{C,0}^2}{R}e^{\frac{-t}{RC}} \tag{1.28}$$

And the energy added to the capacitor by means of this process:

$$E_C = \int_0^\infty P(t)dt \tag{1.29}$$

$$= \frac{(V_{dd}^2 - V_{C,0}^2)}{2}C \tag{1.30}$$

Similarly the total energy delivered by the supply voltage source can be calculated:

$$E_{tot} = V_{dd}(V_{dd} - V_{C,0})C \tag{1.31}$$

The total energy delivered by the supply is higher since power loss occurs due to the resistive losses in R. Thus part of the energy is lost, this is quantified by the charging efficiency $\eta_{C,Charge}$ (Mike Wens and Michiel Steyaert 2011) :

$$\eta_{C,Charge} = \frac{E_C}{E_{tot}} \tag{1.32}$$

$$= \frac{1}{2}\frac{V_{C,0} + V_{dd}}{V_{dd}} \tag{1.33}$$

The previous equation was derived for the charging period to reach infinity, in practice this is a good approximation as long as the voltage settled within 10 % of the charging voltage.

This demonstrates that the efficiency [8] of charging a capacitor only depends on the ratio of initial voltage and the charging voltage. And additionally it is clear that even when no resistance is present at all: a power loss will occur. Thus for capacitive converters to operate in an efficient way, one has to make sure that during operation there is only a small difference between the charging voltage and the initial voltage over the capacitor. But Eq. 1.30 demonstrates that for small differences the amount of energy that is transferred is small as well. This is countered by using a large capacitor.

1.3.5 Inductive Conversion

Inductors

Inductors are passive components storing energy in the magnetic field ($\vec{B}$). This magnetic field can be generated by currents. The inductor is commonly built by winding a conductor around a core with an as high as possible permeability.[9] The inductance L is the ability of an inductor to store energy in a magnetic field and with help of the inductance the voltage across an inductor V_L can be calculated as a function of the instantaneous change in current I through the inductor:

$$V_L = L\frac{dI}{dt} \tag{1.34}$$

To pass a current through an inductor an external source must perform work. The amount of work corresponds to the energy stored in the inductor. The power is expressed as:

$$\frac{dW}{dt} = IV_L \tag{1.35}$$

[8] Theoretically efficiency is expressed in terms of power instead of energy, but it is clear that if taken in a periodical repeating process, and under steady state, time cancels out in the equation.

[9] In integrated processes, an air coil inductor is often used due to the lack of proper core material.

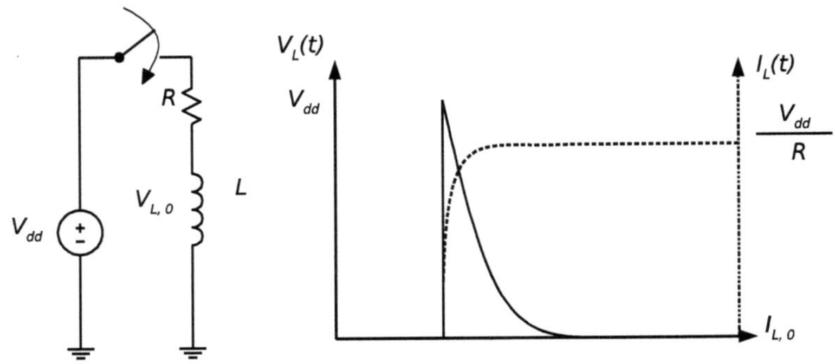

Fig. 1.20 Putting energy in an inductor by changing the current through the inductor

By substituting Eq. 1.34 in 1.35 and integrating from time zero until the current reaches magnitude I, the expression of the energy contained within the inductor (or work performed to achieve this) is:

$$W = \frac{LI^2}{2} \tag{1.36}$$

Energy Transfer

Inductive converters use an inductor to transfer energy from the input to the output. Instead of going deep into the multiple topologies and equations describing these topologies, we have a look at the simple process of storing energy in an inductor.[10] Figure 1.20 represents the current and voltage through and across the inductor, when the switch is closed. This simple process already reveals the key characteristics of using an inductor in a switched-mode inductive converter.

The process is described by the following differential equation:

$$-V_{dd} + RI_L + L\frac{dI_L}{dt} = 0 \tag{1.37}$$

Solving towards V_L and deriving the current through the capacitor,leading to:

$$V_L(t) = (V_{dd} - RI_L(0))e^{\frac{-tR}{L}} \tag{1.38}$$

$$I_L(t) = \frac{V_{dd}}{R} + (I_L(0) - \frac{V_{dd}}{R})e^{\frac{-tR}{L}} \tag{1.39}$$

The power stored in the inductor is:

[10] More information on this topic can be found in (Mike Wens and Michiel Steyaert 2011).

$$P_L(t) = V_L(t)I_L(t) \tag{1.40}$$

$$= \frac{V_{dd}^2}{R}\left(e^{\frac{-tR}{L}} - e^{\frac{-2tR}{L}}\right) \tag{1.41}$$

And the energy inserted into to the inductor by means of this process:

$$E_L = \int_0^\infty P_L(t)dt \tag{1.42}$$

$$= \frac{LV_{dd}^2}{2R^2} \tag{1.43}$$

The charging efficiency calculation is omitted in this work but it can be demonstrated that the η can reach 100 % irrespective of the boundary conditions and given R equals zero. But if R is non zero, look at Eq. 1.43 , the maximum energy that can be stored is limited by the series resistance of the inductor: R.

1.3.6 Analysis

The difference between energy storage and transfer by means of a capacitor and an inductor, is the following. For inductive conversion, current keeps on flowing and resistive losses are permanently present. The period of increasing the inductor current must be halted once the maximum energy content is (nearly) reached. The maximum energy content of an inductor is determined by the resistor in the charging path, this is demonstrated in Eq. 1.43. While for capacitive conversion the resistor will not play a role, as long as the circuit's time constant is much smaller than the observed time period. On the other hand for capacitors the charging efficiency is function of the ratio of the charging voltage and the initial voltage, the latter is demonstrated in Eq. 1.33. This demonstrates that to obtain an as high as possible energy transfer, either large capacitors for a capacitive DC–DC converter or a large inductor with a small series resistance for an inductive DC–DC converter are required.

This discussion is based on the elementary characteristics of the different DC–DC conversion techniques and does not take the practical implementation of such a converter into account. But without doubt this analysis reveals the fundamental bottlenecks of both techniques: Inductive type DC–DC converters look an appealing choice when low-resistance inductors (high-quality factor) are available. For production technologies where capacitors with a high capacitance density are present, capacitive type DC–DC converters are a viable alternative.

1.4 State-of-the-Art Integrated Converters

The miniaturization of power supplies can be approached from two different perspectives: PowerSIP and PowerSoC. PowerSip is a System-in-Package (SIP) approach where as much functionality as possible is integrated on a single IC and the passives are packaged along with the IC. The PowerSOC approach aims for the fully integrated SoC approach both for the control functionality, switches as for the passive components. In the following paragraphs the most relevant integrated converter realizations are demonstrated. This section gives a broad overview of the techniques and specifications of the trend-setting state-of-the-art DC–DC converters.

1.4.1 Inductive Converters

Off-Chip Passives

The first step in the integration of DC–DC converters is the evolution from DC–DC converters built entirely from discrete components towards integration of the switches and control loop on the same die.[11] This approach delivers a first significant reduction in footprint. Up to today this approach is widely investigated especially in applications where an efficiency above 80 % in combination with a small footprint are a must-have. In (Hazucha et al. 2005) a four-phase DC–DC converter utilizing on-package air-core inductors is presented. The high degree of integration permits the shift towards relatively high switching frequencies (up to 233 MHz) with respect to the conventional of-the-shelf DC–DC converters. The high switching frequency enables the use of relatively small passive components which reduces the footprint furthermore. On the other hand the excellent quality of the passive components delivers a solution with an efficiency of 80–87 %. The power density of 213 mW/mm^2 is quite high but does not include the area occupied by the off-chip inductors.

Bond-Wire Inductors

Discrete-type inductors invoke serious limitations on the further reduction of the converter's footprint. Therefore it would be a great advantage to get rid of the separate inductor. It is demonstrated in Craninckx and Steyaert (1996) that for RF-circuits a bond-wire inductor can be used as an alternative for on-chip or discrete-type inductors. The relatively low resistance of such an inductor results in a good quality factor. The latter makes it appropriate to use them in DC–DC converters as well. The use of such a bond-wire inductor in a DC–DC converter is presented in Wens and Steyaert (2007). Omitting the discrete passive components and replacing them by the

[11] Some papers erroneously use the words *fully integrated* or monolithic while partially integrated is obviously a more appropriate term.

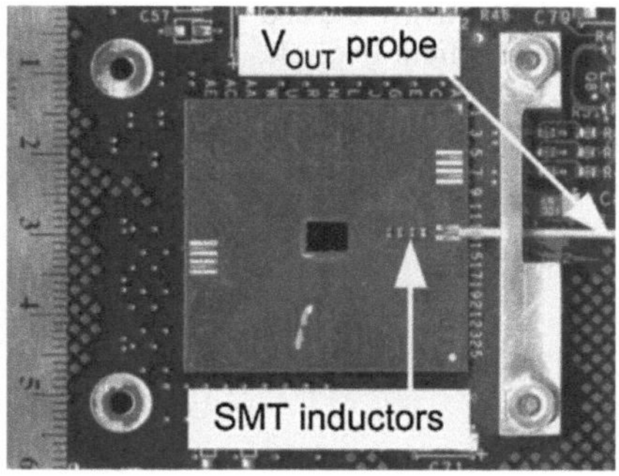

Fig. 1.21 Photograph of the prototype in Hazucha et al. (2005). Four air core inductors are mounted on the package

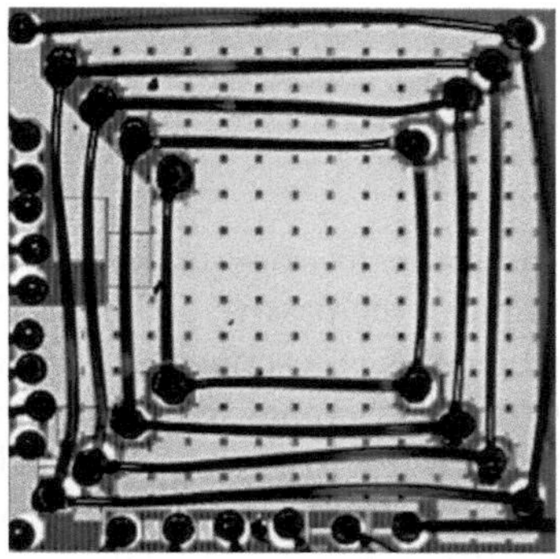

Fig. 1.22 Photograph of the prototype in Wens and Steyaert (2007). The inductor is constructed by means of the spiral of connected bond-wires

bond-wire inductor and an on-chip capacitor provides again a significant reduction in footprint but requires additional die handling and bond-wire manipulation. Although this is a step towards full integration (the output capacitor is entirely integrated on chip), this integration has its impact on the performance of the DC–DC converter. The maximum efficiency is only 65 % with an EEF of 23 % for a maximum output power of 300 mW and a power density of 133 mW/mm^2.

Passive-Die Approach

Monolithic integration of inductors in a standard CMOS process has found to be quite troublesome. But 3D-stacking of dies, exploiting the third dimension is a promising approach to reduce footprint while increasing chip area. This technique can equally be addressed for building DC–DC converters. The most striking advantage of 3D-stacking is that dies of different nature/technology can be connected very closely. The stacked dies are connected by means of solder bumps (Karadi et al. 2008) (the number of dies is then restricted to two) or by through silicon vias (no restriction in number of dies at all).

The active devices are then implemented in CMOS exploiting the excellent switch-characteristics and cost benefit, while the passive devices are integrated in a so-called

Fig. 1.23 Passive die and Active Die from Bergveld et al. (2009)

passive technology that has limited layers and process steps but is optimized to improve the passive component's characteristics. In Bergveld et al. (2009) such a realization is described, the paper describes a DC–DC down converter using a 65 nm CMOS technology to integrate the active die and a dedicated passive die to integrate the inductor. By means of flip-chip technology both dies are connected. While the peak efficiency reaches 87.5 % an EEF of 10 % is demonstrated and a power density 8 mW/mm^2. In Fig. 1.23 the prototype is shown.

Fully Integrated Inductive Converters

This section focuses on plain-CMOS integrated inductive DC–DC converters. Examples exist of prototypes which use additional exotic processing steps to enhance the implementation of inductors. The use of additional steps is discouraged due to the lack of compatibility with the state-of-the-art digital processes. And since digital circuits are the most attractive circuits to be powered by a fully integrated power-management system, all measures should be taken to promote compatibility.

2004 The DC–DC converter in Richelli et al. (2004) is regarded as the first fully integrated inductive DC–DC converter. The prototype is a 1.8 to 6 V boost converter, built in a 0.18 μm technology and achieves 28 % efficiency at a load of 600 μA. The authors of this work clearly point to the lossy inductor as the major source of power loss and the relatively low efficiency.

2007 While it remains quiet around integrated DC–DC converters for a long time, in 2007 a prototype is presented on ISSCC in Alimadadi et al. (2007). This time it is a buck-type converter with a 48 % efficiency and a negative EEF (−8 %), resulting in worse performance than a linear-type regulator. The use of the prototype is motivated by addressing the converter both for DC–DC conversion and feeding a clock-three by charge recycled from the DC–DC converter.

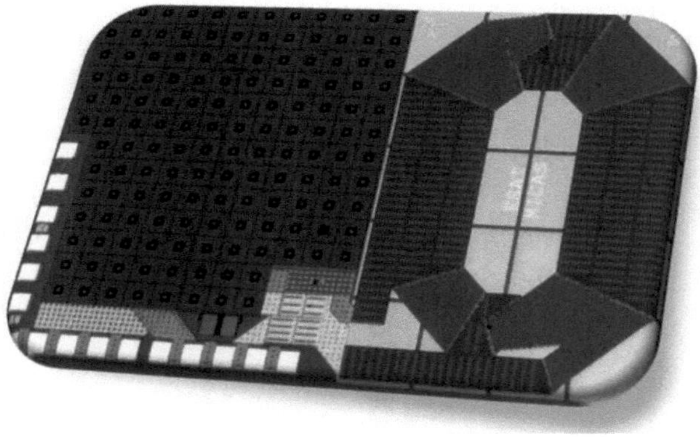

Fig. 1.24 Integrated spiral inductor from Wens et al. 2008

In Wens et al. (2008) a CMOS monolithic DC–DC converter, shown in Fig. 1.24, demonstrating considerable efficiency enhancement is presented . The maximum EEF is 12 % while a power density of 53 mW/mm^2 is achieved and a total output power of 180 mW.

2008–2011 The next few years the absolute efficiency remains the main goal of many designs, 78 % in Wibben and Harjani 2008 and 70 % in Jinhua Ni et al. (2009). Meanwhile the EEF remains modest: respectively 4 and 23 %. In Wens and Steyaert (2011) an 800 mW converter is demonstrated with an EEF of 21 % and a maximum efficiency of 57 %.

1.4.2 Capacitive Converters

Discrete Capacitors

Capacitive DC–DC converters typically require a higher number of components than the equivalent inductive types do. Therefore the capacitive converters using external components are typically topologies with a small number of capacitors: voltage doublers or voltage dividers.

In Lau et al. 2007 a partially integrated voltage doubler is presented with a maximum efficiency of 91 %. The converter, shown in Fig. 1.25, uses 1 μF flying capacitor and a 2.2 μF output buffer capacitor for supplying a maximum of 150 mA at a 3.3 V output. On the chip photograph, the switches $M_{r/i}$ are indicated. Thanks to the large capacitance values of the flying capacitors, this has been achieved at a maximum switching frequency of 500 kHz. The output noise of the converter is reduced by incorporating a linear regulator in the converter-topology. In Gregoire (2006) and Su et al. (2008) other converters with discrete-type capacitors are presented, they achieve respectively 89 and 91.5 % efficiency primarily thanks to the use of high

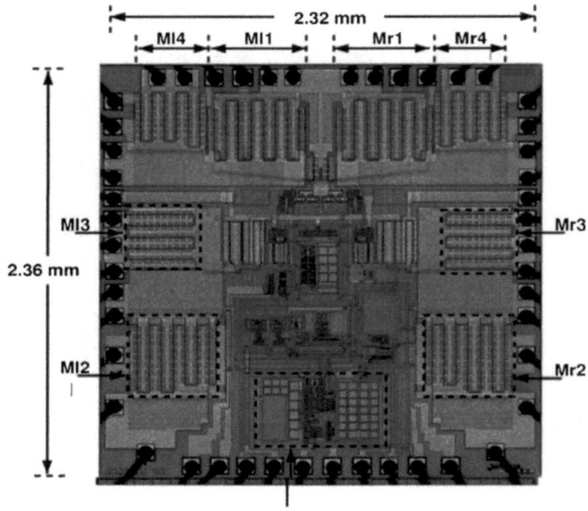

Fig. 1.25 Partially integrated voltage doubler from Lau et al. 2007, only the switches are integrated on-chip

quality external components. Enabling them to reduce the switching losses in the converter.

Fully-Integrated Capacitive Converters

1976 In Dickson (1976) a fully integrated charge pump is presented. The Dickson topology is primarily used for the on-chip generation of high voltages while delivering low current. The Dickson-type converters are used mainly for supplying high voltage for solid-state memories. Figure 1.26 demonstrates the first solid-state implementation of the Dickson converter. As is indicated in the figure, considerable amount of area is dedicated to the on-chip oscillator, the drivers of the switches and the output limiter. The multiplier chain, which is the conversion stage of the converter, is relatively small. Primarily since at a switching frequency of 1 MHz, only 10 μA load current is supplied.

1998 The first fully integrated capacitive converter used in a power-management context is found in Favrat et al. (1998). This paper concerns an improved type of voltage doubler (later also used in Lau et al. 2007), that introduces a number of additional biasing techniques to reduce the charge leakage and latch-up risk.

2008 In Seeman et al. 2008 the use of two separate capacitive DC–DC converters in a power-management IC for a PicoCube sensor node is demonstrated. The converters achieve high efficiencies up to 84.3 % at power levels up to 4 mW partially thanks to the use of high-quality MIM-capacitors. This efficiency is considerably higher than the converters with the same power range and using low-quality capacitors (Minyu et al. 2008).

2010 The domain of medium-power converters (>100mW) was until recently reserved for capacitive converters using external components (Lau et al. 2007; Gre-

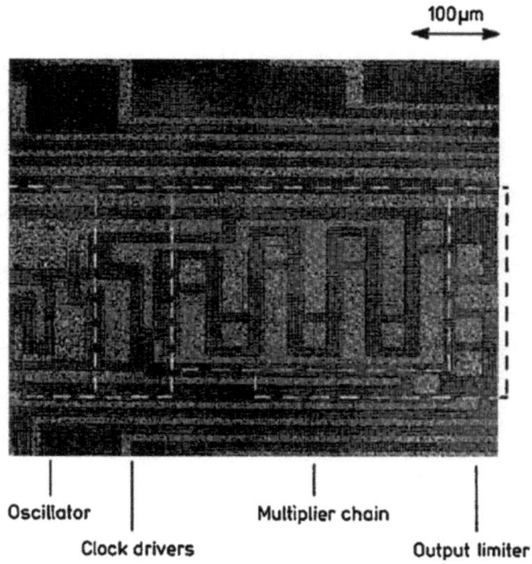

Fig. 1.26 Fully-integrated Dickson Charge Pump from Dickson (1976)

goire 2006; Su et al. 2008). In (Le et al. 2010) an open-loop converter is presented that is implemented in 32 nm Silicon On Insulator (SOI) technology. This converter achieves a power density of 550 mW per square mm and clearly demonstrates the potential of integrated capacitive DC–DC converters. In Chang et al. (2010) a converter is presented that benefits from the extremely high capacitance density of deep-trench capacitors ($200\frac{nF}{mm^2}$) and achieves a current density of 2.3 A per square mm. This high power density is demonstrated by means of an 8.88 mW open-loop capacitive DC–DC converter.

1.4.3 Figures of Merit

To keep track of technological evolutions so-called Figure Of Merits (FOM's) are used to compare circuit performance. In case of power converters: efficiency and power density are the most important static FOM's while line and load regulation characterize the quality of the control system and are dynamic FOM's. The essence of FOM's is that a fair comparison can be made irrespective of the non-relevant features or boundary conditions. On the other hand: FOM's are often abused to put forward excellent features of a design while neglecting the poor ones. In this chapter the EEF was put forward as a more appropriate alternative than comparing efficiency in case of down converters. Therefore, EEF together with power density is put forward as the mayor static FOM's for down-conversion DC–DC converters.

Combining the FOM's in a single equation ignores the non-linearity between power density and EEF. On the other hand the broad range of topologies and structures

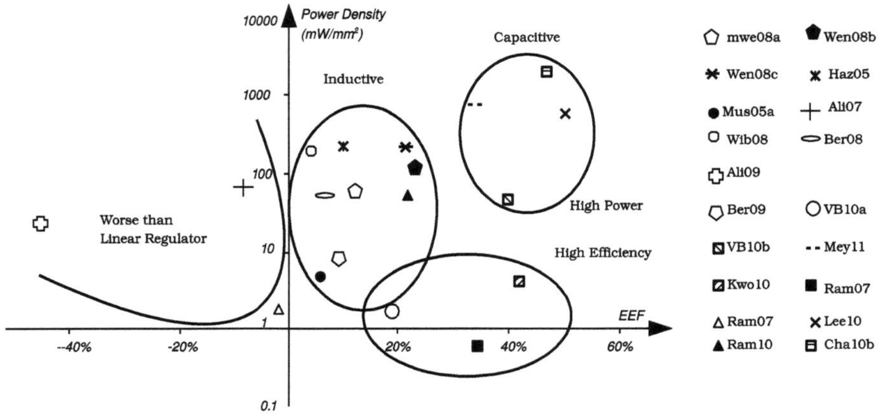

Fig. 1.27 Analysis of the DC–DC converters based on nature of conversion

renders it impossible to equalize the non-linearity. Therefore it makes more sense to compare them visually as is demonstrated in Fig. 1.27.

In Fig. 1.27 an overview is given of the most recently published converters given their power density and EEF. It is clear that there is a broad diversity of converters. The most striking observations are the following. The published inductive converters score particularly high on the power density FOM (Hazucha et al. 2005; Wens et al. 2008; Wibben and Harjani 2008; Karadi et al. 2008) but in general they fail to demonstrate high EEF. The main issue is clearly the lack of high-quality inductors in CMOS. The capacitive converters are divided in two groups. The first group is aiming for high EEF but has low power density (Breussegem and Steyaert 2010; Chandrakasan and Ramadass 2007). The second group is scoring particularly good on both FOM's (Le et al. 2010; Chang et al. 2010). In Fig. 1.28 a more in-depth look is taken on these capacitive converters. The division between the two groups of capacitive converters corresponds to the division between technology options. While the bulk CMOS DC–DC converters achieve high EEF with relatively low power density, except for Meyvaert et al. (2011). The exceptionally good performing converters are built in Deep Sub Micron (DSM) technologies (Le et al. 2010; Chang et al. 2010), exploiting the extreme dense and high-quality capacitors available in that technology.

1.4.4 Analysis

Capacitive DC–DC converters have been around for a long time (Evennat and Lorrain 1953; Dickson 1976) but it was believed that their application area was restricted to low-current applications and thus they were perceived to be inferior to the more versatile inductive converters. Indeed capacitive DC–DC converters are bounded

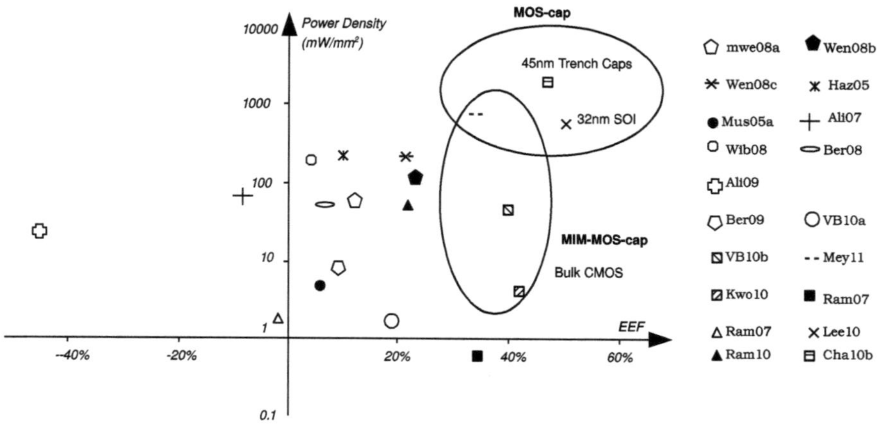

Fig. 1.28 Analysis of capacitive DC–DC converters based on technology

to a single optimal VCR which is determined by the topology. Next to this the component count is typically higher than it is for inductive DC–DC converters. Since extra components need extra bond-wire connections to the driver IC's, the solution becomes more expensive. Thanks to the migration to integrated DC–DC converters both these obstacles were removed.

During the last couple of years integration of capacitive DC–DC converters has been demonstrated primarily for low-power solutions. High efficiency can be achieved particularly thanks to the use of high-quality MIM-caps. This efficiency is considerable higher than the converters with the same power range and using low-quality capacitors. The domain of medium-power converters ($>100\,mW$) was until recently reserved for integrated inductive converters and capacitive converters using external components primarily thanks to the use of high quality external components. Le et al. (2010) clearly demonstrates the potential of SOI technology for integration of capacitive DC–DC converters. In Chang et al. (2010) a converter is presented that benefits from the extremely high capacitance density of deep-trench capacitors.

The research of fully integrated capacitive DC–DC converters has mainly focused on low power (sub-mW) applications. While a lot of portable consumer applications (mW-W-range) can benefit even more from an efficient integrated DC–DC converter. The recent developments are focusing on increasing power density by using advanced technologies (SOI—Deep Trench), while a lot of applications require compatibility with circuits in cheap standard CMOS. The control aspect of these type of converters is neglected while it is of primary importance to ensure proper operation. Therefore there is a clear need for research on these topics.

1.5 Summary and Outline

This book aims for introducing mixed-signal electronic designers with the concepts of integrating capacitive DC–DC converters in CMOS. Additionally a number of new ideas and techniques developed throughout the previous years are highlighted. The chapters are organized as follows:

- Chapter 1: describes the context of this work. Introduces the reader with the concepts of System-on-Chip (SoC), power management, the different power-management techniques and the need for on-chip DC–DC conversion.
- Chapter 2: offers a first look at the operation of capacitive DC–DC converters and gives a broad overview of the dominantly used capacitive converter topologies.
- Chapter 3: builds a model for capacitive DC–DC converters and presents a design flow called Output Impedance Balancing. The Output Impedance Balancing Technique is extended for multi-objective or multi-topology capacitive converter design.
- Chapter 4: Introduces a number of existing and novel techniques developed to reduce the noise generation in fully integrated capacitive DC–DC converters.
- Chapter 5: presents an overview of existing and newly developed control techniques. These techniques are tightly matched to the integration-specific characteristics of the converters.
- Chapter 6: goes deeper into the nature of the CMOS process and how the CMOS features can be exploited to improve the capacitive converter's performance.
- Chapter 7: presents a number of prototypes and goes into detail on how the techniques in previous chapters are deployed.

1.6 Conclusion

This chapter has shown that there are three main drivers for developing high-performance, small-footprint DC–DC converters. First the increasing power density of state-of-the-art digital circuits is posing problems. Next the voltage gap between battery voltages and the CMOS technology maximum supply voltages is raising serious issues. And finally the energy gap between consumer expectations and battery chemistry, requires effective techniques to reduce power consumption in state-of-the-art applications.

A number of techniques are being explored to solve the power-related problems: ranging from clock gating to voltage scaling and body biasing. Many of these techniques require an interface between the actual power supply and the loading circuit. This circuit should perform a DC to DC conversion and do this in the most efficient way possible while meeting a whole range of requirements (accuracy, bandwidth,...). Linear series regulators can be very useful but demonstrate low efficiency when used for small VCR's.

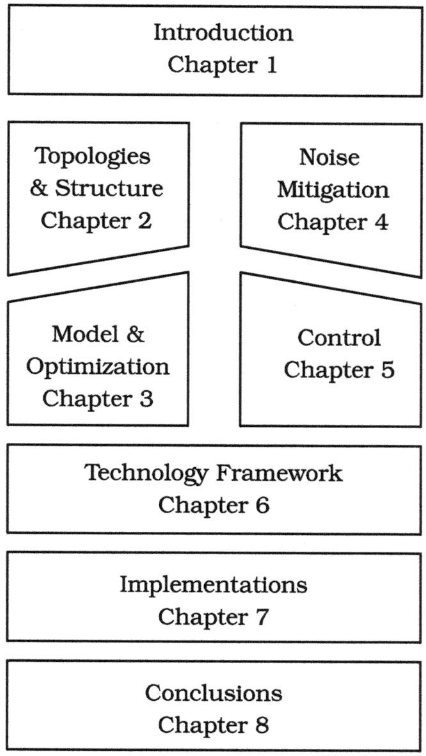

Fig. 1.29 Graphical Representation of the table of contents and the relationship between the chapter

Both inductive and capacitive type converters provide a potentially more efficient solution. Analysis of the energy transfer in both types of converters suggest that fully integrated inductive converters struggle with the quality factor of the inductor and this point of view is supported by the most recent prototypes published in scientific literature. A capacitive converter's efficiency is less vulnerable to the parasitic resistance and this suggests that using capacitive conversion instead is a viable approach. The following chapters will demonstrate which methodologies can be used for achieving high efficiency and high power density by means of this technique.

Chapter 2
Converter Topologies and Fundamentals

To build a solid understanding of the capacitive conversion technique, this chapter introduces the fundamental characteristics of capacitive DC–DC converters in Sect. 2.1. Section 2.2 discusses three different analysis techniques: Charge Flow Analysis, Charge Balance Analysis and Branch Analysis. These complementary techniques have their specific merit in the development of the state-of-the-art capacitive DC–DC converter modeling techniques (Chap. 3) and are used for analyzing capacitive DC–DC converters. Charge Flow Analysis demonstrates an intuitive method to determine the VCR of capacitive converters. Charge Balance Analysis builds a bridge between the conventional switched-capacitor analysis and the analysis of capacitive DC–DC converters. Branch Analysis presents a generalized technique to analyze and to qualify the topology performance. Next the converter taxonomy is discussed in Sect. 2.3 and finally the analysis techniques are demonstrated for a selection of capacitive DC–DC converter topologies in Sect. 2.4.

2.1 Characteristics

DC–DC conversion by means of capacitors differs fundamentally from an inductive DC–DC converter. The most notable difference is that lossless conversion can only be achieved at infinitely high switching frequencies or by a converter with an infinitely large amount of capacitance. In practice a properly designed capacitive DC–DC converter faces only a small efficiency penalty for violating these requirements. To make this more tangible, this section offers a first look into the principles and the operation of a primitive capacitive DC–DC converter.

2.1.1 DC–DC Converter Structure

A capacitive DC–DC converter consists of the two parts in Fig. 2.1: the conversion block and the control block. The conversion block is the heart of the converter and

T. Van Breussegem and M. Steyaert, *CMOS Integrated Capacitive DC–DC Converters*, Analog Circuits and Signal Processing, DOI: 10.1007/978-1-4614-4280-6_2,

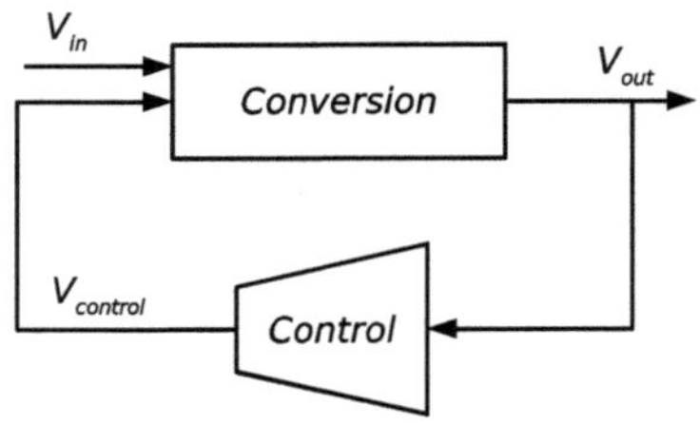

Fig. 2.1 Graphical representation of the coarse converter system partitioning: the conversion block performs the DC–DC conversion and the control block controls the behavior of the conversion block

performs the actual conversion between the DC input voltage and the DC output voltage. The control block is a signal processing system that manipulates the behavior of the conversion block in order to keep it in line with the system requirements, it is discussed in Chap. 5.

There is a clear difference in nature between both blocks. The conversion block is the so-called power stage and embodies the low impedance part of the DC–DC converter. The control part is a high impedance feedback path of the conversion characteristics (output voltage, output current,...) to at least one of the control parameters of the conversion block (switching frequency, duty cycle of the switching frequency,...).

A DC–DC converter can operate without a closed control loop. This is done in case there is no power budget for the control loop circuitry or in case that the behavior of the DC–DC converter is non critical or has large design margins (Max1682, switched capacitor voltage doubler). This is rarely done, since the conversion characteristics are typically very sensitive to variations in operation circumstances (varying load, varying input voltage, temperature). The latter is demonstrated in Chap. 3.

2.1.2 Principles

The capacitive DC–DC converter is identified as a Variable Structure System (VSS). The operation of such a VSS is characterized by the repetitive change in the circuit's structure. In contrast to inductive converters, capacitive converters use only switches and capacitors to transfer charge between the input and the output. In fact the capacitive converter consists of two distinct types of capacitors: the flying capacitors and the output buffer capacitor.

The flying capacitors are the charge-transferring capacitors. While the buffer capacitor does not participate in the charge transfer, it mainly influences the start-up behavior of the converter and the steady-state noise characteristics. In general the periodical reconfiguration of the switched-capacitor structure exists of two or more states.[1]

[1] In practice multi-state converters are a rarity (Ben-Yaakov and Kushnerov 2009). Two-state converters are used in most of the publications.

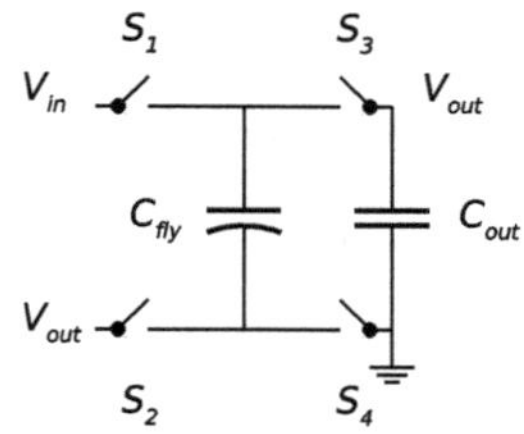

Fig. 2.2 The conversion block of a fractional $\frac{1}{2}$ converter: it consists of the flying capacitor C_{fly}, the output buffer capacitor C_{out} and four switches S_i

One of the most important constraints faced during the design of a capacitive DC–DC converter is the following. Each converter topology has an ideal VCR (iVCR). This iVCR is the maximum ratio between the output voltage and the input voltage of the conversion block. In practice this iVCR is the upper bound for the actual VCR and the converter can only operate at a theoretical efficiency of 100 % when this VCR is met. For a converter with an input voltage V_{in}, an output voltage V_{out} and a common ground connection. The actual VCR is defined as:

$$VCR = \frac{V_{out}}{V_{in}} \tag{2.1}$$

A certain topology corresponds to a single iVCR, but a given iVCR can be obtained by multiple topologies. This gives the designer a range of possible implementations for achieving a certain conversion or conversion range and this is discussed in Sect. 2.3. In order to substantiate the previous description, the operation of capacitive DC–DC converters is demonstrated by means of the most straightforward capacitive DC–DC converter: the series-parallel $\frac{1}{2}$ converter.

2.1.3 Example: The Series-Parallel $\frac{1}{2}$ Converter

The series-parallel $\frac{1}{2}$ converter consists of one flying capacitor C_{fly} and an output buffer capacitor C_{out}. The terminals of the flying capacitors undergo a significant change in potential due to the periodic change in converter structure. The output buffer capacitor does not participate in the charge transfer related to the conversion, it only reduces the switching noise that originates from the switched nature of the converter. One of the output capacitor's terminals is connected to ground or another DC voltage in the circuit. In Fig. 2.2 a series-parallel $\frac{1}{2}$ converter topology is shown.

The two-state operation of a series-parallel $\frac{1}{2}$ converter is formed by alternating between the following configurations:

$$\phi_1 : S_1 = S_2 = 1, S_3 = S_4 = 0 \tag{2.2}$$

$$\phi_2 : S_1 = S_2 = 0, S_3 = S_4 = 1 \tag{2.3}$$

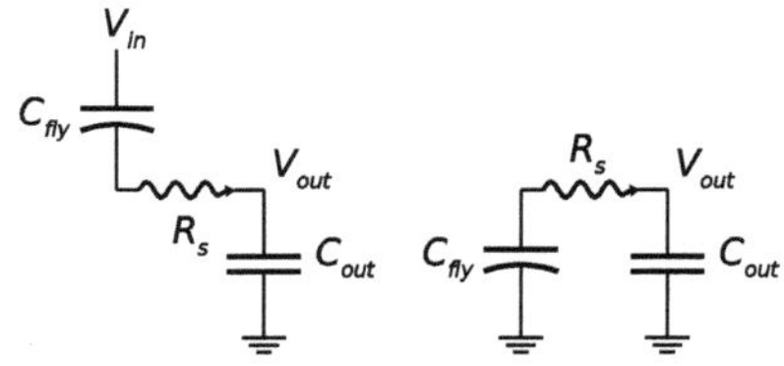

Fig. 2.3 Graphical representation of the two states of a fractional $\frac{1}{2}$ converter: the *left pane*: state ϕ_1 and the *right pane*: state ϕ_2

Those two states are graphically represented in Fig. 2.3. The left-pane structure corresponds to state ϕ_1 and the right pane to state ϕ_2. By alternating in a periodical way the steady state voltage at the output ideally corresponds to $\frac{V_{in}}{2}$.

In order to demonstrate the circuit's behavior, the system's set of time-dependent differential equations is formulated. For sake of generality an equivalent series resistance R_s is included. This resistance can be either invoked by the non-zero on-resistance of the switches, by the series resistance of the capacitors or a combination of both.

For state ϕ_1:

$$V_{out} = V_{in} - V_{C_{fly}} - R_s C_{fly} \frac{dV_{C_{fly}}}{dt} \tag{2.4}$$

$$C_{fly} \frac{dV_{C_{fly}}}{dt} - C_{out} \frac{dV_{out}}{dt} = 0 \tag{2.5}$$

For state ϕ_2:

$$V_{out} = V_{C_{fly}} + R_s C_{fly} \frac{dV_{C_{fly}}}{dt} \tag{2.6}$$

$$C_{fly} \frac{dV_{C_{fly}}}{dt} + C_{out} \frac{dV_{out}}{dt} = 0 \tag{2.7}$$

By solving this system of differential equations of state ϕ_2 for the boundary conditions $V_{C_{fly},0} = V_{C_{fly},-}$ and $V_{C_{out},0} = V_{C_{out},-}$ the output voltage can be obtained as a function of time (t):

$$\begin{aligned} V_{out} = & \frac{C_{fly} V_{C_{fly},-} + C_{out} V_{out,-}}{C_{fly} + C_{out}} \\ & - C_{fly} \frac{V_{C_{fly},-} - V_{out,-}}{C_{fly} + C_{out}} e^{-\frac{C_{out}+C_{fly}}{R_s C_{fly} C_{out}} t} \end{aligned} \tag{2.8}$$

$$I(C_{out}) = C_{out} \frac{dV_{out}}{dt} = \frac{V_{C_{fly},-} - V_{out,-}}{R_s} e^{-\frac{C_{out}+C_{fly}}{R_s C_{fly} C_{out}} t} \tag{2.9}$$

The peak current is observed at $t = 0$:

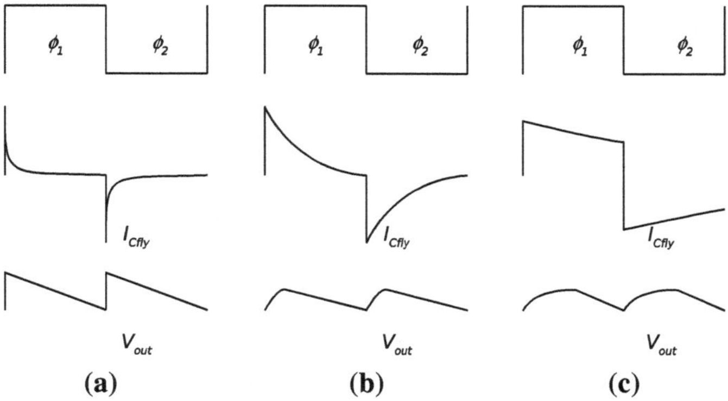

Fig. 2.4 **a** Full Charging Mode (FCM), **b** Boundary Charging Mode (BCM), **c** Partial Charging Mode (PCM)

$$I_{C_{fly},+,peak} = \frac{V_{C_{fly},-} - V_{out,-}}{R_s} \tag{2.10}$$

The current pattern in Eq. 2.9 is dominated by an exponential decay, the rate is determined by the parasitic resistance R_s and the capacitor sizes. While the peak current during changing state of this Fractional Converter is only proportional to the voltage difference before changing state and inversely proportional to the parasitic resistance in the circuit R_s. Notice that if no resistance is present the charge transfer finds place by means of a current pulse ($I_{C_{fly},+,peak} = \infty$) and the decay time turns to zero.

Based on the time constants of the charge transfer, early literature (Zhu and Ioinovici 1996) describes three operation modes: The Full Charging Mode where the parasitic resistance is relatively small and the flying capacitor current decays to zero during each state or switching interval. The Boundary Charging Mode for which the current decay corresponds to half of the switching period and the Partial Charging Mode where the current pattern resembles a linear slope and the flying capacitor current does not turn to zero during commutation. This effect of the parasitic resistances can be observed in Fig. 2.4. Negligible parasitic resistance is observed when the charge transfer resembles a current pulse in Fig. 2.4a, large parasitic resistance in Fig. 2.4c is identified by the quasi-constant current pattern.

2.2 Analysis Techniques

In this section three analysis techniques are selected: Charge Flow Analysis, Charge Balance Analysis and Branch Analysis. The selected techniques demonstrate a comprehensive introduction to the most prominent modeling approach for capacitive DC–DC converters: The Output Impedance Model.

2.2.1 Charge Flow Analysis

Charge Flow Analysis is the primary tool for identifying the role of the different components in the conversion block. Based on this analysis, the charge flow vectors a_c^i are extracted. These vectors play an important role in the modeling and design techniques presented in this work. They qualify the capacitive converter performance and enable an objective comparison of the converter topologies.

A conversion block consists of a number of linear networks switched periodically to achieve the charge transfer (Wu and Bass 2000). A first step in the analysis of such an array is to identify the component configurations for the separate states of the conversion and to formally describe the converter topology. Therefore a set vectors is defined (Seeman and Sanders 2008) describing the topology based on the charge flow through the components. $q_i^{(j)}$ represents the amount of charge that is transferred during state j by capacitor i. q_{out} is the total amount of charge transferred to the load during a switching period T.

The charge flow through the capacitors is described by:

$$a_c^{(1)} = [q_{out}^{(1)} \;\; q_1^{(1)} \;\; \ldots \;\; q_n^{(1)} \;\; q_{in}^{(1)}] \;/\; q_{out} \tag{2.11}$$

$$a_c^{(2)} = [q_{out}^{(2)} \;\; q_1^{(2)} \;\; \ldots \;\; q_n^{(2)} \;\; q_{in}^{(2)}] \;/\; q_{out} \tag{2.12}$$

These charge vector elements can be determined by inspection for every state of the conversion period based on the following principles:

- Kirchhoff's current law in each node: The sum of charge flow elements equals zero in each circuit node.
- In steady state, for every component the sum of both state's charge flow elements equals zero.
- The output capacitor C_{out} is much larger than the flying capacitors and behaves as a voltage source with respect to the remainder of the circuit. This assumes no voltage ripple at the output node.

This is demonstrated by an analysis of a $\frac{1}{2}$ series-parallel converter in Fig. 2.5. For this converter:

$$a_c^{(1)} = [q_{out}^{(1)} q_{C_{fly}}^{(1)} q_{out}^{(1)}]$$

$$a_c^{(1)} = [\frac{1}{2} \;\; \frac{1}{2} \;\; \frac{1}{2}] \tag{2.13}$$

$$a_c^{(2)} = [\frac{1}{2} \;\; \frac{-1}{2} \;\; \frac{0}{2}] \tag{2.14}$$

The following observations are made. The charge vector elements of both states of the flying capacitors have opposite signs. The charge vector elements of the output sum to 1. The ratio of the total input and output charge vector elements equals the iVCR (N) of the topology:

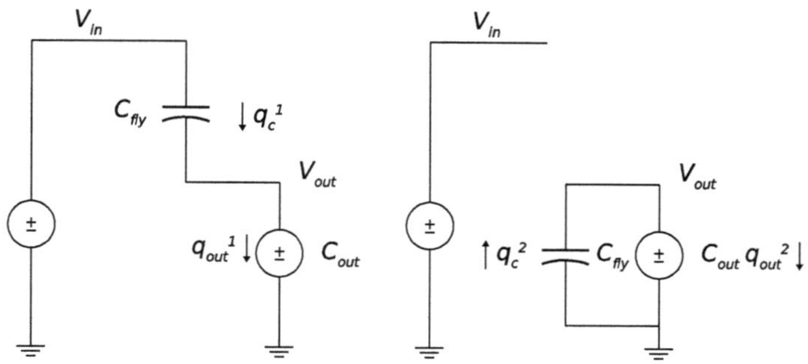

Fig. 2.5 Charge Balance Analysis of a fractional $\frac{1}{2}$ converter: the *left pane*: state ϕ_1 and the *right pane*: state ϕ_2

$$N = \frac{q_{in}}{q_{out}} \tag{2.15}$$

This demonstrates that this intuitive method can be used for determining the converter's iVCR.

2.2.2 Charge Balance Analysis

Charge Balance Analysis is used in the analysis of switched-capacitor circuits and is based on the law of charge conservation (Tsividis 1983). Charge Balance Analysis deals with the absolute amount of charge in the circuit, this in contrast with the Charge Flow Analysis that concerns the change in charge on the components. Therefore the charge flow analysis which is depicted in Fig. 2.6 uses the quantity q_x and the charge balance method uses the quantity Q_x. These quantities have the same dimension, Coulomb, but q_x denotes a change in Q_x for a change in time $\Delta t = t_1 - t_0$:

$$q_x = Q_{x,t=t_1} - Q_{x,t=t_0} \tag{2.16}$$

In steady state the charge is conserved along both states of the switching period and the output voltage is constant:

$$\Sigma Q_{c,i}^{(1)} = Q_{out}^{(2)} + \Sigma Q_{c,i}^{(2)} \tag{2.17}$$

The charge stored on the capacitors (including on the output capacitor) is calculated based on the voltage-capacitance-charge relationship. The charge dissipated in the load has been determined by Ohm's law.

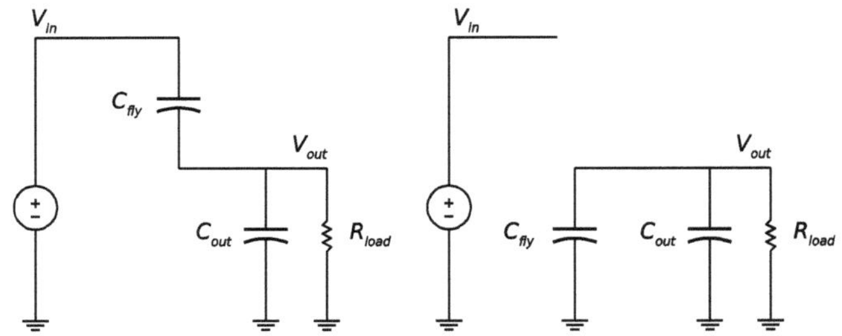

Fig. 2.6 Discrete Time Analysis of a fractional $\frac{1}{2}$ converter: the *left pane*: state ϕ_1 and the *right pane*: state ϕ_2

$$Q_{out}^{(2)} = \frac{V_{out}[n-1]}{2R_{load}f_{sw}} \tag{2.18}$$

This results in the following charge conservation equation:

$$\begin{aligned} &C_{fly}(V_i[n-1] - V_{out}[n-1]) + C_{out}V_{out}[n-1] \\ &\quad = \frac{V_{out}[n]}{2R_{load}f_{sw}} + C_{fly}V_{out}[n] + C_{out}V_{out}[n] \end{aligned} \tag{2.19}$$

This Discrete Time equation can be transformed into the circuit's transfer function (TF) by means of the Z-transform (Dorf 1995):

$$\frac{V_{out}}{V_{in}} = \frac{C_{fly}z^{-1}}{(C_{fly} + C_{out} + \frac{1}{2R_{load}f_{sw}}) - (-C_{fly} + C_{out})z^{-1}} \tag{2.20}$$

This transfer function demonstrates the input-output voltage relationships in function of the switching frequency f_{sw} the capacitor sizes C_{out}, C_{fly} and the load R_{load}.
First the transfer function 2.20 is analyzed for the unloaded case ($R_{load} = \infty$) and in case no AC variation ($z = 1$):

$$\frac{V_{out}}{V_{in}} = \frac{C_{fly}}{(C_{fly} + C_{out}) - (C_{out} - C_{fly})} = \frac{1}{2} \tag{2.21}$$

For the loaded converter: transfer function 2.20 is evaluated in $z = 1$ and this gives:

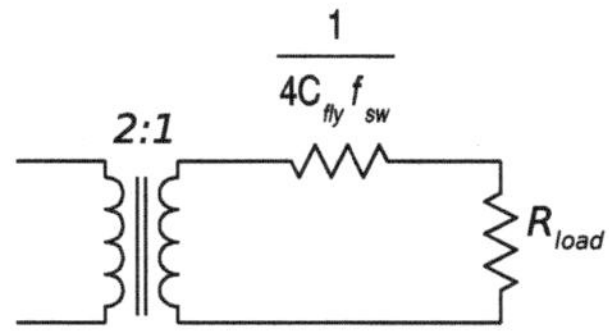

Fig. 2.7 Schematic representation of the equivalent circuit found by applying the Charge Balance Method to a series-parallel $\frac{1}{2}$ capacitive converter

$$\frac{V_{out}}{V_{in}} = \frac{C_{fly}}{\left(C_{fly} + C_{out} + \frac{1}{2R_{load}f_{sw}}\right) - (-C_{fly} + C_{out})} \tag{2.22}$$

$$= \frac{1}{\left(2 + \frac{1}{2C_{fly}R_{load}f_{sw}}\right)} \tag{2.23}$$

Equation 2.23 can be simplified:

$$V_{out} = \frac{1}{2}V_{in}\frac{R_{load}}{(R_{load} + \frac{1}{4C_{fly}f_{sw}})}. \tag{2.24}$$

This input–output relationship consists of two factors: the constant gain factor $\frac{1}{2}$ and the load dependent gain factor $\frac{R_{load}}{(R_{load}+\frac{1}{4C_{fly}f_{sw}})}$. Actually considering a constant C_{fly} and f_{sw}, this gain equals the gain of a resistive divider. The same transfer function can be achieved by cascading an ideal DC–DC transformer with a fixed gain $\frac{1}{2}$ and a resistive divider. This is demonstrated in Fig. 2.7. This means that a capacitive converter is determined by a conversion ratio N and an output impedance. Analysis of other topologies by means of this Charge Balance Method results in similar solutions.

2.2.3 Branch Analysis

The Charge Balance Analysis is a technique to analyze a capacitive conversion block but for complex converter topologies it turns to be an exhaustive method. Therefore the following analysis has been developed to determine the output impedance, based on Tellegen's theorem: Branch Analysis. The Branch Analysis involves two approximations of the output impedance. The first includes nothing but the effect of the switched-capacitor nature of the converter. It is described by the Slow Switching Approximation. The second approximation only includes the resistive nature of the converter and is described by the Fast Switching Approximation. First Tellegen's theorem is revisited:

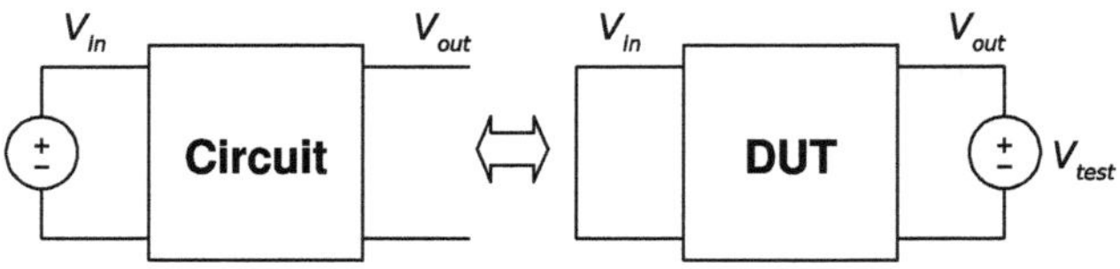

Fig. 2.8 **a** Regular set-up of the converter. **b** The converter as a Device Under Test in a set-up to determine the output impedance

Tellegen 1 *Consider an arbitrary lumped network whose graph G has b branches and n_t nodes. Suppose that to each branch of the graph we assign arbitrarily a branch potential difference W_k and a branch current F_k for $k = 1, 2, \ldots, b$, and suppose that they are measured with respect to arbitrarily picked associated reference directions. If the branch potential differences $W_1, W_2, \ldots, W_b$ satisfy all the constraints imposed by KVL and if the branch currents $F_1, F_2, \ldots, F_b$ satisfy all the constraints imposed by KCL, then*

$$\Sigma_1^b W_k F_k = 0 \qquad (2.25)$$

The Slow Switching Approximation

The Charge Balance analysis suggests that a capacitive converter can be modeled as a voltage-dependent voltage source (or a DC-transformer) with a non-zero output impedance. The most straightforward technique to determine the output impedance of a circuit is to apply a test source at the output terminals of the circuit and to short circuit the input of the circuit. This is demonstrated in Fig. 2.8

Tellegen's theorem can be interpreted as follows: for each state the charge balance vectors are orthogonal with the voltages across the components. Superposition of both states leads to the following equality:

$$v_{out}(a_{out}^{(1)} + a_{out}^{(2)}) + \Sigma_{i=1}^{n}(a_{c,i}^{(1)} v_{c,i}^{(1)} + a_{c,i}^{(2)} v_{c,i}^{(2)}) = 0 \qquad (2.26)$$

The previous equation can be simplified by taking into account that $a_{out}^{(1)} + a_{out}^{(2)} = 1$. Moreover the absolute value of the charge balance vector elements and the voltage difference over the capacitors is introduced: $a_{c,i} = a_{c,i}^{(1)} = -a_{c,i}^{(2)}$ and $q_i = a_{c,i} q_{out}$. These simplifications reduce Eq. 2.26 into:

$$v_{out} q_{out} + \Sigma_{i=1}^{n}(q_i \Delta v_i) = 0 \qquad (2.27)$$

If it is assumed that the capacitors are linear or demonstrate a linear behavior in the operation point they are used. Then $\Delta v_i = \frac{q_i}{C_i}$. By dividing Eq. 2.27 by q_{out}^2 the following equality is obtained:

$$\frac{v_{out}}{q_{out}} - \Sigma_{i=1}^{n} \left(\frac{q_i}{q_{out}}\right)^2 \frac{1}{C_i} = 0 \tag{2.28}$$

If both terms are divided by the switching frequency f_{sw} the following formulation is obtained:

$$\frac{v_{out}}{q_{out} f_{sw}} = \frac{v_{out}}{i_{out}} = \Sigma_{i=1}^{n} \left(\frac{q_i}{q_{out}}\right)^2 \frac{1}{f_{sw} C_i} \tag{2.29}$$

By definition $a_{c,i} = \frac{q_i}{q_{out}}$, then it is readily derived that the output impedance corresponds to the ratio of the test voltage and the resulting current. Moreover the output impedance is only a function of the amount of flying capacitance, the switching frequency and the sizing of the flying capacitors:

$$R_{SSL} = \Sigma_{i=1}^{n} \frac{a_{c,i}^2}{f_{sw} C_i} \tag{2.30}$$

For this derivation the influence of the parasitic resistors in the circuit are ignored and therefore this output impedance is only valid considering no influence of the resistance in the circuit.

The Fast Switching Approximation

If the parasitic resistance can not be ignored and the power loss due to these resistances is dominant, another approach is followed. A set of switch charge flow vectors is determined. This can be done based on the same methodology proposed in the Charge Flow Analysis. There are two switch charge vectors each corresponding to each one of the converter's states. Each element $a_{r,i}$ of the vector corresponds to the charge flow through one of the switches S_r.

The average current through the switches equals the amount of charge divided by the commutation period $DT = \frac{D}{f_{SW}}$:

$$i_{r,i} = \frac{q_{r,i} f_{sw}}{D} \tag{2.31}$$

Considering that $q_{r,i} = a_{r,i}q_{out}$ and $q_{out} = \frac{i_{out}}{f_{sw}}$ and the charge flow of a two-state capacitive converter is optimum for dutycycle $D = 0.5$:

$$i_{r,i} = 2a_{r,i}i_{out} \tag{2.32}$$

The power loss in the switches can now be formulated as:

$$P_{loss,switches} = \Sigma\left(\frac{1}{2}R_i(2a_{r,i}i_{out})^2\right) \tag{2.33}$$

Since the power loss is proportional to the square of the output current, the output impedance is:

$$\frac{P_{loss,switches}}{i_{out}^2} = \Sigma\left(\frac{1}{2}R_i 2a_{r,i}^2\right) \tag{2.34}$$

$$R_{FSL} = 2\Sigma_i R_i a_{r,i}^2 \tag{2.35}$$

So the R_{FSL} is the output impedance, if the losses in the switches are dominating the total losses in the capacitive DC–DC converter. This approach inevitably leads to a dual interpretation of a capacitive DC–DC converter's output impedance. Each of these interpretations has nothing in common with the other one: For the capacitive nature the resistance in the circuit is ignored, for the resistive nature, the switched capacitor nature is ignored. Both approaches can be unified by considering both natures as complementary. In Seeman and Sanders (2008), it is demonstrated that the total output impedance is accurately approximated as follows:

$$R_{out} = \sqrt{R_{SSL}^2 + R_{FSL}^2} \tag{2.36}$$

2.3 Topologies: Taxonomy

The topology represents the spatial properties of the capacitive DC–DC converter. More specifically: the structure and interconnection of the converter's components are described. This can be done either by a literal description, for example the netlist of a circuit or by a graphical means, for example a schematic drawing of the circuit components. Since capacitive converters are variable-structure systems with different states, it is preferred to represent each state by a separate schematic. This method has been used in previous sections.

The previous sections dealing with the analysis techniques indicate that each capacitive converter topology has a distinct iVCR. This fundamental property of capacitive DC–DC converters has large repercussions on the use of these converters. First, a single topology has an upper bound for which conversion can be performed:

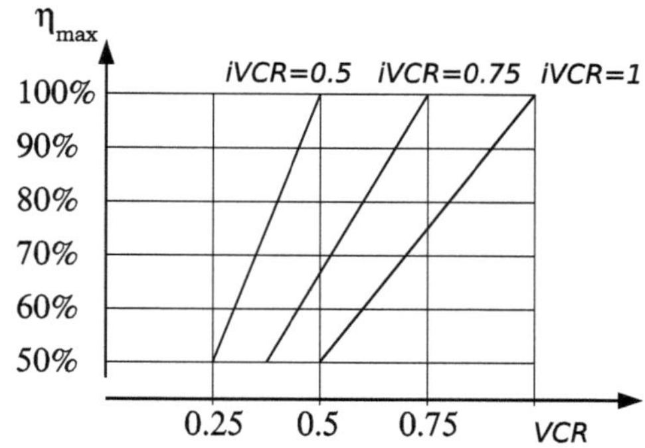

Fig. 2.9 In the graph the maximum efficiency with respect to the VCR is demonstrated for three different topologies

the iVCR. Secondly the efficiency has an upper bound that corresponds to the ratio of the actual VCR and the iVCR:

$$VCR \leq iVCR \tag{2.37}$$

$$\eta_{max} = \frac{VCR}{iVCR} \tag{2.38}$$

This relationship is represented in Fig. 2.9. This demonstrates that the maximum attainable efficiency heavily decreases for deviations from the iVCR. This effect puts constraints on the input-output voltage range of a capacitive converter topology if high efficiency is of the designer's concern. Therefore the VCR is one of the primary aspects of a capacitive DC–DC converter and classification of the capacitive converter topologies is required.

2.3.1 Topology Occurrence Theorem

There exists a vast range of converter topologies. The occurrence of capacitive converter topologies is subject to an important theorem (Makowski and Maksimovic 1995), which predicts the achievable iVCR given a certain number of capacitors:

The theoretical occurrence of capacitive type DC–DC converter topologies is defined as follows: For a two-state capacitive type of DC–DC converter with n flying capacitors, the flying capacitors can be configured such that an ideal VCR N is achieved.

$$N(n) = \frac{V(n_{out})}{V(n_{in})} = \frac{P[n]}{Q[n]} \tag{2.39}$$

The VCR N_i not only corresponds to the ideal ratio of the output voltage and the input voltage but also corresponds to the ratio between two integer numbers $P[n]$ and $Q[n]$. These characteristic numbers satisfy the following inequalities:

$$Max[Abs[P[n]], Abs[Q[n]]] \leq F_n Min[Abs[P[n]], Abs[Q[n]]] \geq 1 \tag{2.40}$$

In these equations N is the ideal VCR, n the number of flying capacitors and F_n the n-th Fibonacci number. If the number of flying capacitors is limited to three, the following conversion ratios can be achieved:

n flying capacitors	N_i
1	$\frac{1}{2}, 1, 2$
2	$\frac{1}{3},\frac{1}{2},\frac{2}{3},\frac{3}{2},1,2,3$
3	$\frac{1}{5},\frac{1}{4},\frac{1}{3},\frac{2}{5},\frac{1}{2},\frac{3}{5},\frac{2}{3},\frac{3}{4},\frac{4}{5},1,\frac{5}{4},\frac{4}{3},\frac{3}{2},\frac{5}{3},2,\frac{5}{2},3,4,5$

2.3.2 Up Converters

Historically seen, the use of capacitive DC–DC converters focused on up converters. Up-converters have a VCR larger than one. In the early 1900s these converters were used to generate high voltages (in the range of kV) in particle physics experiments. But the break-through of the solid-state transistor renewed the attention for voltage multipliers. By the 1970s capacitive DC–DC converters (charge pumps) found their way as integrated voltage generators in the memory business.[2] It appears that the research interest in up converters has been tempered over the past years, especially due to the emerging potential of down converters and the maturity of the modified Dickson Charge Pumps (Mensi et al. 2005).

[2] Embedded memories require voltage of 5–10 V which is typically higher than the supply voltage of the accompanying chips.

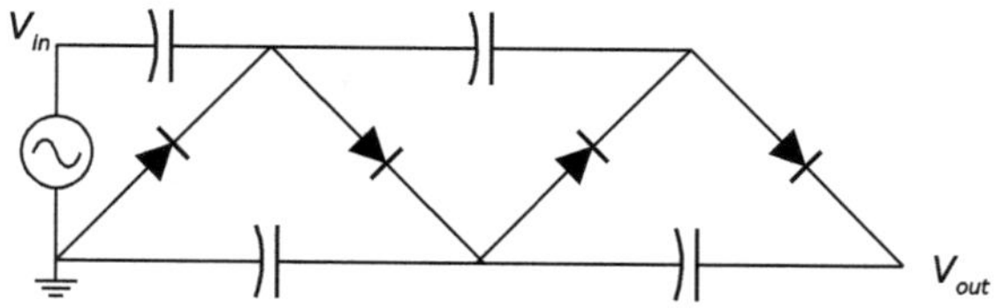

Fig. 2.10 Greinacher topology

The Greinacher Multiplier

The Greinacher multiplier was invented in 1914 by Heinrich Greinacher, a Swiss physicist (Evennat and Lorrain 1953). It is actually an AC-DC converter but since it served as inspiration for one of the most used DC–DC converters (Dickson 1976) it deserves some attention. In Fig. 2.10 a two-stage Greinacher converter is shown. The physicists John D. Cockroft and Ernest T.S. Walton used this converter to generate extremely high voltages (>100 kV) for their particle physics experiments and this converter became also known as the Cockroft–Walton Voltage Multiplier. The strength of this converter lays within its simplicity: it requires nothing but a number of diodes and capacitors, and no active timing circuitry nor switches.

It operates at follows: during the negative half wave the upper branch capacitors are charged to the peak input voltage. During the positive half wave, the lower branch capacitors are charged to two times the peak input voltage. After a start-up period, the output voltage reaches steady state and an output voltage that becomes ideally equal to $2nV_{peak}(n_{in})$, n being the number of stages. Moreover the components are only facing a voltage corresponding to two times the peak voltage of the AC-input, while the converter is able to generate voltages that are 2n times higher than the peak input voltage. This concept is used to deal with high voltages while the individual components of the system are facing a small fraction of the high voltage is nowadays known as voltage domain stacking.

> This voltage domain stacking is actually a concept used in some of the most state-of-the-art DC–DC converters in Deep Sub Micron CMOS: if the individual components have limited voltage capability, a topology is used that exposes the components to only a fraction of the total voltage (Van Breussegem and Steyaert 2011; Somasekhar et al. 2010; Le et al. 2010; Ng et al. 2009).

The Dickson Charge Pump

The Dickson Converter or Charge Pump finds its origin in the need for on-chip high-voltage generation (high with respect to the conventional on-chip voltages: >10 V). These high voltages are required for erasing and writing the non-volatile solid-state memories. This invoked a renewed interest in voltage multipliers and for the first time a need for on-chip voltage multipliers. The main bottleneck of the Greinacher

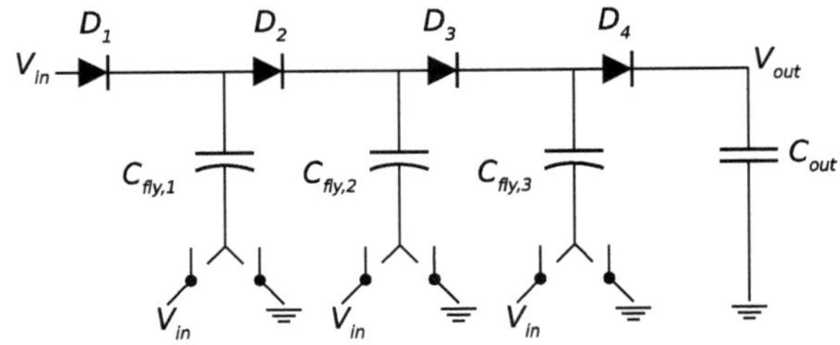

Fig. 2.11 Dickson topology

Multiplier laid within the series connection of the capacitors. Practical capacitors and especially on-chip capacitors exhibit a large parasitic capacitance to the chip's substrate (stray capacitance) and the Greinacher Voltage Multiplier proves to be very sensitive to these parasitic elements. Especially due to the series connection of the capacitors. Therefore a new topology (Dickson 1976) was introduced in 1976, that demonstrates a structure based on parallel connection of the capacitors instead of series connection. This topology demonstrates a much lower output impedance and a much higher resilience to stray capacitance. In Fig. 2.11 a three-stage Dickson is represented. An ideal unloaded Dickson converter with n stages, n equally sized flying capacitors with a total flying capacitance C_{fly}, a stray capacitance α, switching at a frequency f_{sw} and loaded with a current I_{load}, has the output voltage

$$V_{out} = V_{in} + \frac{n}{1+\alpha}\left(V_{in} - \frac{I_{load}}{f_{sw}C_{fly}}\right) \tag{2.41}$$

The Dickson charge pump operates as shown in Fig. 2.12: In a first state the odd numbered diodes in the diode string conduct and transfer charges in the direction of the load through the diode string. In the second state, the even-numbered diodes conduct and the odd-numbered diodes block. This sequence is invoked by alternating the potential of the flying capacitors bottom plate by means of the switches. In more advanced implementations (Mensi et al. 2005) the diodes are replaced by active devices and feed-forward biasing[3] techniques are adopted to decrease the threshold voltage and thus the drop-out voltage over the switches in the previously called 'diode string'.

This charge pump has the same advantages as the Greinacher Multiplier concerning the diodes and switches: these are exposed to a fraction of the output voltage. But the capacitors are dealing with much larger voltages, but a lot of capacitor types have a large breakdown voltage by nature and this is exploited in the following design: Van Breussegem and Steyaert (2011). In this converter,

[3] Feed-forward biasing denotes the biasing of diode-connected MOS devices by means of a next stage voltage node in the diode string, operating at a higher voltage. By doing this the threshold voltage can be reduced and power loss in the diode is cut.

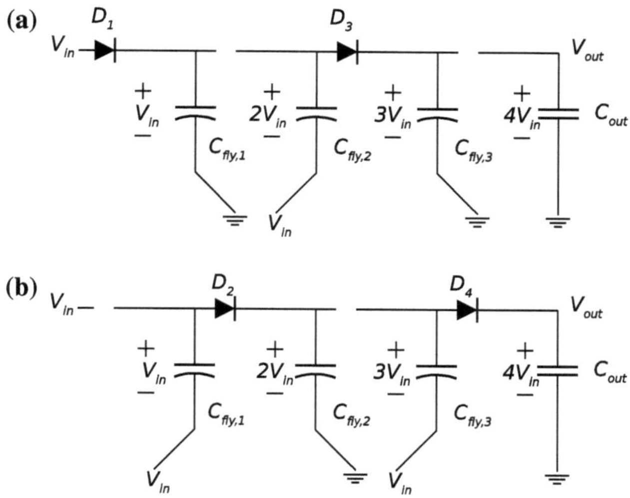

Fig. 2.12 The operation of a Dickson Charge Pump: **a** State 1, **b** State 2

only the capacitors in the level shifter are dealing with larger voltages. The capacitors used herein can cope with voltages up to 10 V.

Parallel-Series Converter

The most straightforward capacitive DC–DC converter is the Parallel-Series Converter. In a first state the flying capacitors are parallel charged by the input source, in the second state the capacitors are series discharged between input and output node. An integer $(n+1)$ Parallel Series converter has n flying capacitors and an ideal voltage multiplication of $(n + 1)$.

The integer $\frac{3}{1}$ Parallel-Series Converter, that is represented in Fig. 2.13, operates in a two-state cycle. During the first state both flying capacitors are charged up to the input voltage. During this state of the commutation the output capacitor is decoupled from the charge-transferring structure. Next the flying capacitors are series-connected between input and output node and in steady state the output voltage equals three times the input voltage, in case no load is applied.

2.3.3 Down Converters

The interest in capacitive DC–DC down converters has grown recently. Especially the so-called *Voltage Gap* discussed in Chap. 1 has pushed capacitive down-converters

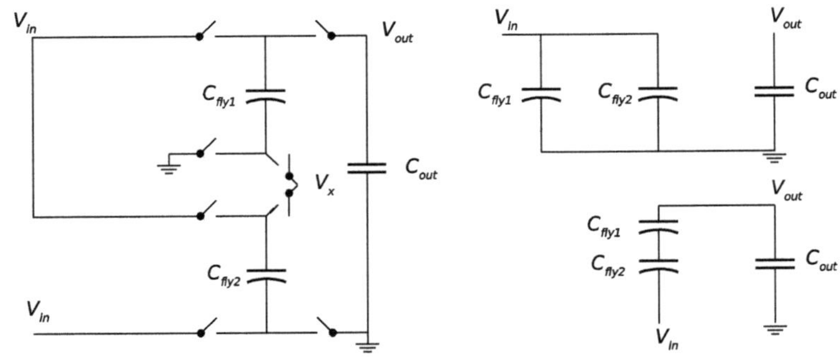

Fig. 2.13 Parallel series converter topology with schematic representation of both configuration states

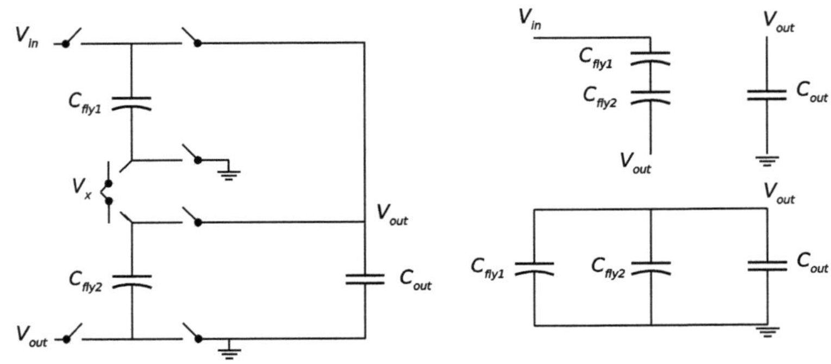

Fig. 2.14 Series-parallel converter topology with schematic representation of both configuration states

into the main stream of power-management solutions. This section describes a number of the dominant down-conversion topologies, found in literature.

Series-Parallel Converter

The Series-Parallel converter is the antagonist of the Parallel-Series Converter. In a first state the flying capacitors are connected in series between input node and output node while in the second state the capacitors are discharged in parallel with the output..A $(n+1)^{-1}$ Parallel-Series converter has n flying capacitors and an ideal voltage multiplication of $(n + 1)^{-1}$. It is actually the mirrored version of the up converter but with transposed input and output nodes.

The $\frac{1}{3}$ capacitive converter is shown in Fig. 2.14, the left pane shows the entire topology while the right pane highlights the configurations during both states of the topology.

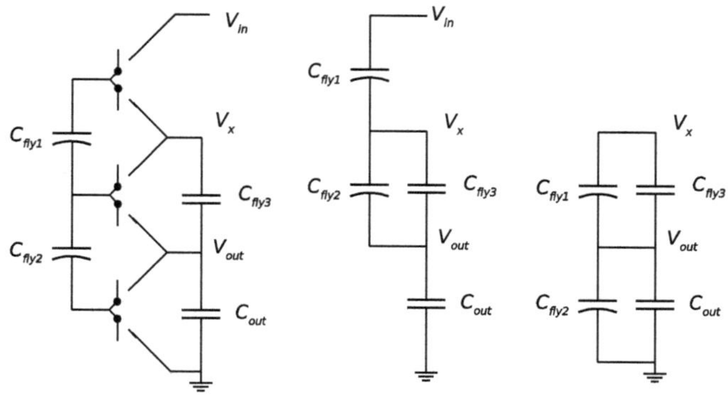

Fig. 2.15 Ladder topology

Ladder Converter

The ladder converter consists of two series-capacitor-strings that slide along each other while charging from the supply and discharging towards the load. Figure 2.15 shows the two states of a ladder converter. A converter with n flying capacitors performs a primary conversion with a ratio $\frac{2}{n+3}$, with n the number of flying capacitors. For a ladder converter each capacitor will charge until a voltage equal to $\frac{(n+3)V_{in}}{2}$ is observed across each capacitor in case no load is applied of course.

The converter has the advantage that a DC voltage can be tapped from multiple nodes (n_{out} n_x for the example). It is thus a multiple-output converter by nature. But loading of multiple nodes will increase the output impedance and thus negatively influence the converter efficiency if the other specifications remain the same. Moreover a type of voltage-domain stacking can be implemented because of the multiple DC nodes in the circuit and this makes this kind of converter very appealing to use in high-input-voltage applications. On the other hand, this type of converter requires a relatively large number of switches and this turns the converter in a switch-intensive solution.

Fractional Converter

The Fractional Converters is a family of converters that cannot be classified under the previous types (Makowski and Maksimovic 1995). In most cases the iVCR of this type is hard to be determined by visual inspection. Formal determination of the iVCR is performed by means of the analysis techniques presented in the previous section. The existence of this kind of converter is predicted by the theorems in Makowski and Maksimovic (1995), but their synthesis is non-methodological. In Fig. 2.16 a $\frac{4}{5}$ fractional converter topology is shown. At the right the component configuration is demonstrated for both conversion states.

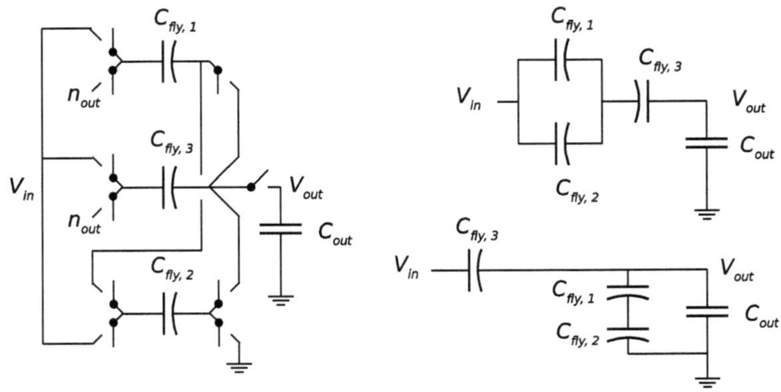

Fig. 2.16 A fractional 4/5 topology and converter states

Fig. 2.17 Graphical representation of the potential efficiency improvement by using the multi-topology approach

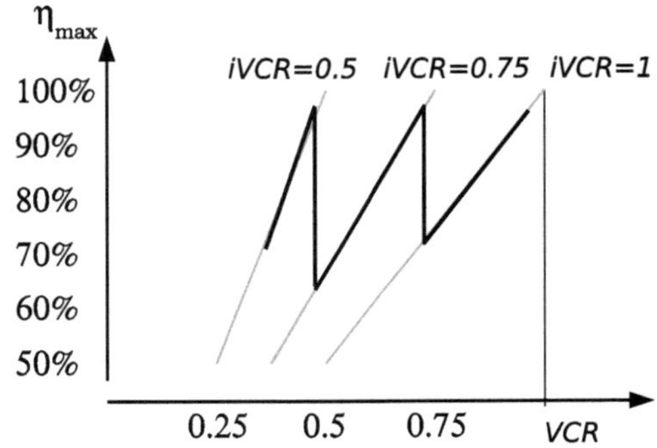

2.3.4 Multi-Topology Converters

The relationship between topology and iVCR puts have constraints on the input-output voltage range of the converter and the associated converter's performance within this range. To improve the efficiency/performance over a broad range and thus to increase the flexibility of the converter (the ability to deal with a broad range of conversion scenario's), there is a clear need for multi-topology capacitive converters.

These converters comprise a capacitor-switch array that can not only switch between both states of the base topology but can also switch between different topologies. Each one of these topologies addresses a separate part of the input-output range. The latter technique is demonstrated in Fig. 2.17. The ideal efficiency ratings of three topologies, based on Eq. 2.38, are drawn in gray. Using either one of them in a separate configuration will either constraint the input output range: the topologies have a maximum VCR corresponding to the iVCR. Or each topology demonstrates a poor performance (low maximum efficiency η_{max}) if the VCR deviates significantly from iVCR but by combining multiple topologies in a single structure and to switch

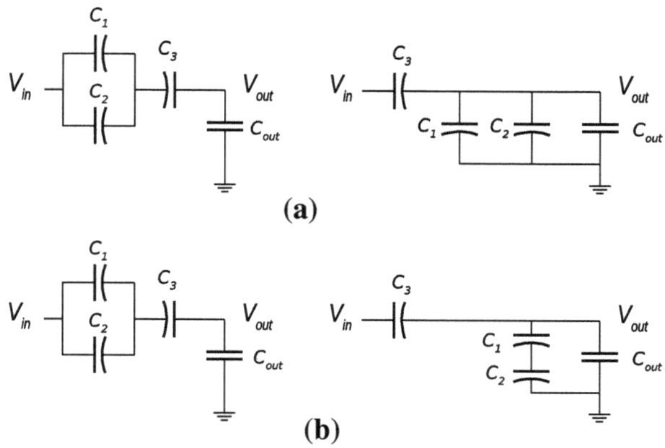

Fig. 2.18 Multi topology converter: **a** the 2/3-topology, **b** the 4/5-topology

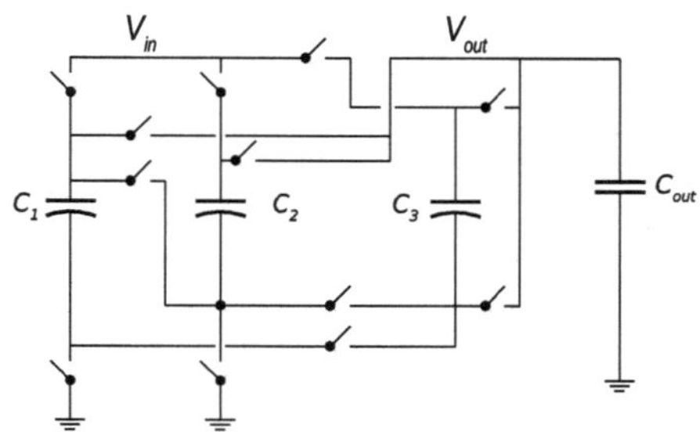

Fig. 2.19 Implementation of a multi topology converter with both a 4/5 and a 2/3 topology

structure according to the required VCR. In Fig. 2.17 the potential maximum efficiency of a multi-topology converter comprising of three topologies (iVCR=$\frac{1}{1}$ $\frac{3}{4}$ $\frac{1}{2}$) is demonstrated by means of the thick dark line. This shows that the input–output range is extended and that in the low VCR range the efficiency is boosted with respect to the single topology approach if only iVCR $\frac{3}{4}$ is used. In Fig. 2.18 the converter's states are shown. In the upper pane (a) the $\frac{2}{3}$- topology, in the lower pane (b) the $\frac{4}{5}$-topology.

In Fig. 2.19 an implementation example is demonstrated of a multi-topology converter comprising of a dual $\frac{4}{5}$ and a $\frac{2}{3}$ capacitive converter.

2.4 Topologies: Analysis

The previous section introduced a selection of topologies occurring in the state-of-the-art capacitive converters. In this section the analysis technique presented before are applied to these topologies. This analysis provides a first look at the topology performance and the opportunities laying herein. For each converter topology, both states are graphically represented including the parasitic switch resistance. Other

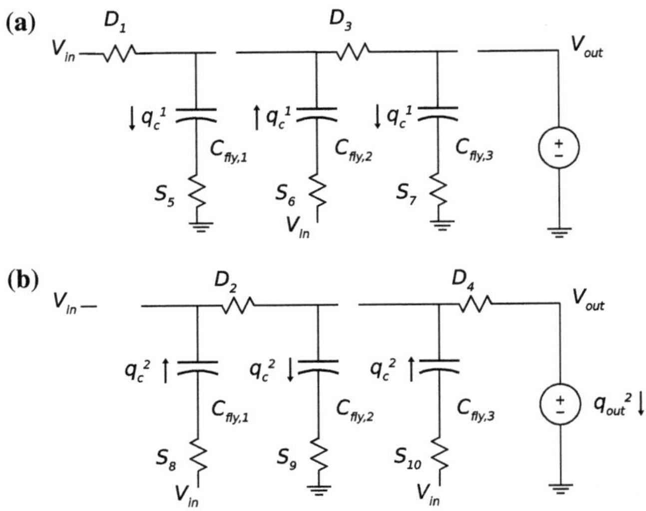

Fig. 2.20 Charge Flow Analysis of the Dickson Converter

parasitic components, for example the parasitic series resistance of the switches or the parasitic resistance related to the metal interconnect between the components, can be included in a similar fashion. But for sake of clarity this is omitted in this first analysis.

2.4.1 Dickson Converter

The Dickson converter is a two state converter (Dickson 1976; Zhang and Llaser 2004). Figure 2.20 represents both states of the converter. In the state-of-the-art implementations of the Dickson Converter (Mensi et al. 2005), the diodes in the diode string are replaced by active switches. During state ϕ_1 switches D_1 D_3 S_5 S_6 S_7 are conducting while the other switches are off. During state ϕ_2 switches D_2 D_4 S_8 S_9 S_{10} are conducting, while the other switches are off.

Based on the charge Flow Analysis the following capacitor charge vectors are derived:

$$\overrightarrow{a_c^{(1)}} = [0(-1)(+1)(-1)(2)]$$
$$\overrightarrow{a_c^{(2)}} = [1(+1)(-1)(+1)(2)]$$

Similarly the switch charge vectors can be calculated:

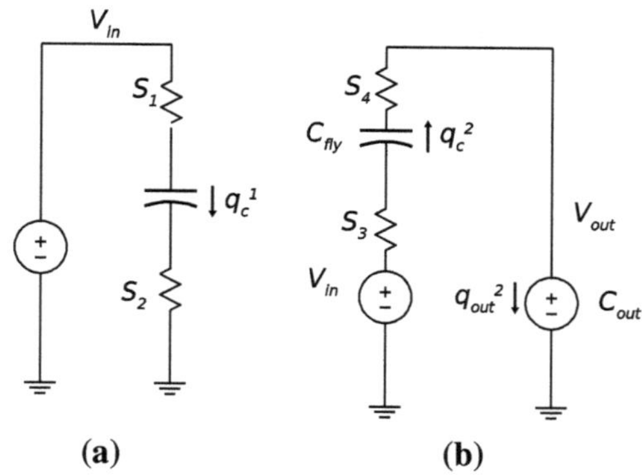

Fig. 2.21 Charge Flow Analysis of the Voltage Doubler

$$\overrightarrow{a_r^{(1)}} = [010101010102]$$
$$\overrightarrow{a_r^{(2)}} = [101010101012] \tag{2.42}$$

For this Dickson converter:

$$N = \frac{q_{in}}{q_{out}} = \frac{2+2}{1} = 4 \tag{2.43}$$

In case that the output impedance is maximized by matching the component sizing with the charge vector elements the following output impedance factors are obtained:

$$R_{SSL} = \frac{(1+1+1)^2}{C_{fly} f_{sw}} \tag{2.44}$$

$$R_{FSL} = 2\frac{(1+1+1+1+1+1+1+1+1+1)^2}{G_{tot}} \tag{2.45}$$

2.4.2 Voltage Doubler

The voltage doubler is a broadly used parallel-series capacitive DC–DC converter (Lee et al. 2006; Gregoire 2006; Lau et al. 2007; Van Breussegem and Steyaert 2009). It is often used to supply a higher voltage at a small part of a chip to improve its performance. In other configurations, multiple voltage doublers are cascaded to achieve a voltage multiplication 2^k with k the number of conversion stages (Starzyk et al. 2001). This results in a larger voltage multiplication than a Dickson converter where the gain increases linearly instead of exponentially. In Fig. 2.21 both states of a single stage voltage doubler topology are shown. The voltage doubler consists of a single flying capacitor C_{fly}, a single output buffer capacitor C_{out} and four switchesS_{1-4}.

The following charge flow vectors are derived for a single stage voltage doubler:

$$\overrightarrow{a_r^{(1)}} = [011001]$$
$$\overrightarrow{a_r^{(2)}} = [100111]$$
$$\overrightarrow{a_c^{(1)}} = [0(-1)(1)]$$
$$\overrightarrow{a_c^{(2)}} = [1(+1)(1)]$$

For this voltage doubler DC–DC converter:

$$N = \frac{q_{in}}{q_{out}} = \frac{2}{1} = 2 \quad (2.46)$$

In case that the output impedance is maximized by matching the component sizing with the charge vector elements the following output impedance factors are obtained:

$$R_{SSL} = \frac{1}{C_{fly} f_{sw}} \quad (2.47)$$

$$R_{FSL} = 2\frac{(1+1+1+1)^2}{G_{tot}} \quad (2.48)$$

2.4.3 Voltage Divider

The voltage divider is the antagonist of the voltage doubler, it is a series-parallel converter. From an output impedance point of view this is the most advantageous converter to implement. Therefore it is often used as a demonstrator circuit to show case the impact of technology improvements on capacitive converter design (Chang et al. 2010; Le et al. 2010). The voltage divider, shown in Fig. 2.22, also consists of a single flying capacitor C_{fly}, a single output buffer capacitor C_{out} and four switchesS_{1-4}. The charge flow vectors are:

$$\overrightarrow{a_r^{(1)}} = \left[\frac{1}{2}, \frac{1}{2}, \frac{1}{2}, 0, 0, \frac{1}{2}\right]$$
$$\overrightarrow{a_r^{(2)}} = \left[\frac{1}{2}, 0, 0\frac{1}{2}, \frac{1}{2}, 0\right]$$
$$\overrightarrow{a_c^{(1)}} = \left[\frac{1}{2}, \frac{-1}{2}, \frac{1}{2}\right]$$
$$\overrightarrow{a_c^{(2)}} = \left[\frac{1}{2}, \frac{1}{2}, \frac{0}{2}\right]$$

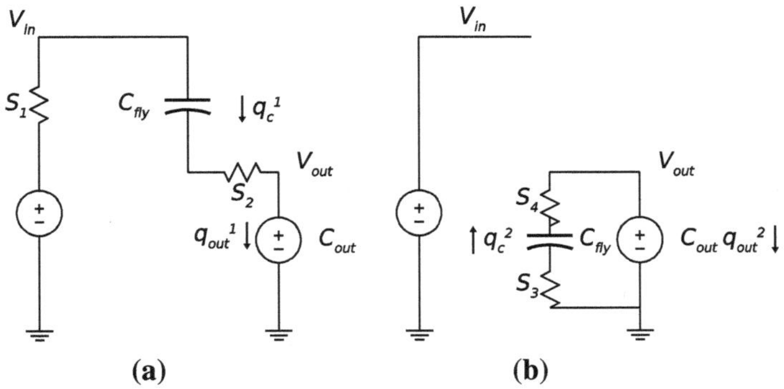

Fig. 2.22 Charge Flow Analysis of the Voltage Divider

For this voltage divider DC–DC converter:

$$N = \frac{q_{in}}{q_{out}} = \frac{1}{2} = 0.5 \tag{2.49}$$

In case that the output impedance is maximized by matching the component sizing with the charge vector elements the following output impedance factors are obtained:

$$R_{SSL} = \frac{1}{4C_{fly} f_{sw}} \tag{2.50}$$

$$R_{FSL} = 2\frac{\left(4 \times \frac{1}{2}\right)^2}{G_{tot}} \tag{2.51}$$

2.4.4 Fractional Converter

Fractional Converters exist for a whole range of conversion ratio's. In practice the number of capacitors is kept below four since the output impedance increases fast in function of the number of capacitors. In Fig. 2.23 both states of a fractional $\frac{4}{5}$ capacitive converter are demonstrated. The topology requires ten switches.

The Charge Flow vectors are:

$$\overrightarrow{a_r^{(1)}} = \left[\frac{1}{5}, \frac{1}{5}, \frac{1}{5}, \frac{1}{5}, \frac{1}{5}, \frac{2}{5}, 0, 0, 0, 0, 0, \frac{2}{5}\right]$$

$$\overrightarrow{a_r^{(2)}} = \left[\frac{4}{5}, 0, 0, 0, 0, 0, \frac{2}{5}, \frac{2}{5}, \frac{1}{5}, \frac{1}{5}, \frac{1}{5}, \frac{2}{5}\right]$$

$$\overrightarrow{a_c^{(1)}} = \left[\frac{1}{5}, \frac{1}{5}, \frac{1}{5}, \frac{-2}{5}, \frac{2}{5}\right]$$

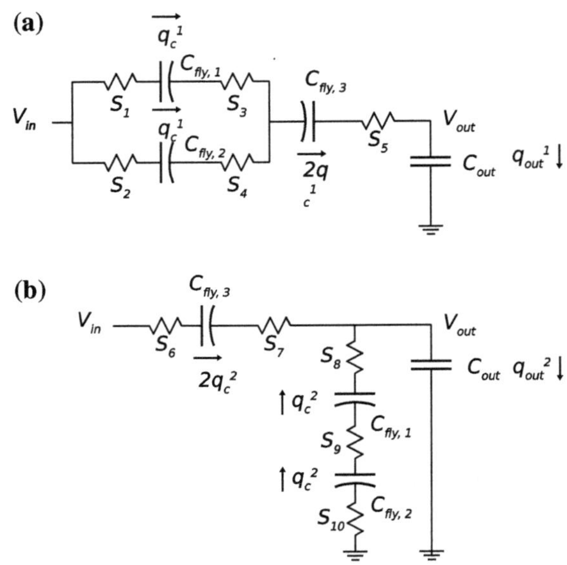

Fig. 2.23 Charge Flow Analysis of the $\frac{4}{5}$ Fractional Converter

$$\vec{a_c^{(2)}} = \left[\frac{4}{5}, \frac{-1}{5}, \frac{-1}{5}, \frac{2}{5}, \frac{2}{5}\right]$$

2.5 Conclusion

This chapter has presented a first look at an alternative DC–DC conversion technique: capacitive DC–DC conversion. This technique distinguishes itself from the conventional techniques by omitting the use of an inductor for achieving the DC–DC conversion. A number of techniques has been elaborated to analyze the operation of the converters and to conduct a first comparison between the different capacitive converter topologies. The predominant capacitive converter topologies are presented and some of their appealing characteristics are mentioned. Both the Dickson Converter and the Greinacher Multiplier use a voltage-domain stacking technique to overcome the technology restrictions of the their components. Moreover the Dickson converter shows that capacitors can be used to bridge the voltage gap between multiple voltage domains.

Now that the operation of the capacitive converter is made clear and a first primitive model (the output impedance model) is suggested, it is time to elaborate this model and to construct a design method for capacitive DC–DC converters.

Chapter 3
Modeling and Design of Capacitive DC–DC Converters

Chapter 2 demonstrates a broad selection of converter topologies and converter types. Additionally a number of converter analysis techniques are presented. These techniques are used to perform a first feasibility analysis for inclusion of capacitive converters in the power-management interface of an integrated system. For example by calculating the maximum theoretical efficiency (Equation 2.38) that can be achieved by the converter topology. This coarse approximation will not match the eventual efficiency of the converter but gives a first impression of the potential of a topology with respect to the alternative topologies. The difference in eventual system performance is determined by the parasitic elements invoked by the physical realization in a (CMOS) technology. Therefore it is quintessential to identify the dominant parasitic effects in the circuit and to model them properly. Additionally the model must be used to maximize the performance of the DC–DC converter. In Chap. 1 a number of performance metrics are enumerated. For a certain operation point, to achieve the highest efficiency given a maximum power density, is the ultimate goal in the domain of fully integrated converters.

This chapter discusses a new methodology for designing capacitive DC–DC converters. Section 3.1 covers the Output Impedance Model. This model serves as a starting-point for the description of the converter operation and is elaborated further in this chapter. Next an intuitive optimization technique, called Output Impedance Balancing (OIB), is introduced in Sect. 3.2. This design technique can be addressed for designing any type of single-topology converter disregarding the technology or the requirements. Next the approach is extended for multi-topology converters in Sect. 3.3. Finally in Sect. 3.4 the Output Impedance Model is adapted to include the effect of a finite-size output capacitor.

3.1 Output Impedance Model

The different analysis techniques in the previous chapter promote the Output Impedance Model as an obvious model for designing a capacitive DC–DC converter. In Fig. 3.1 the most straightforward implementation of this model is represented.

T. Van Breussegem and M. Steyaert, *CMOS Integrated Capacitive DC–DC Converters*, Analog Circuits and Signal Processing, DOI: 10.1007/978-1-4614-4280-6_3,

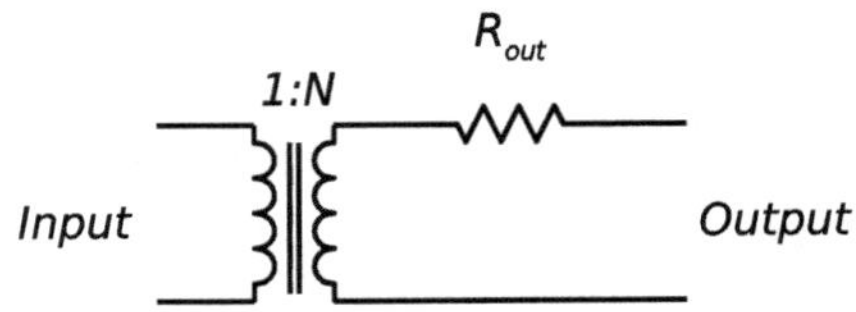

Fig. 3.1 Schematic representation of the output impedance model

The model consists of a DC-transformer with a fixed iVCR N and with a non-zero output impedance R_{out}.

The DC transformer model is a mathematical model to represent an ideal DC–DC converter (RWE01). The two-port model includes the following relationships:

$$V_{out} = N V_{in}$$
$$I_{in} = N I_{out}$$

and is represented by the following two symbols. In capacitive DC–DC converter literature the right-hand representation is preferred.

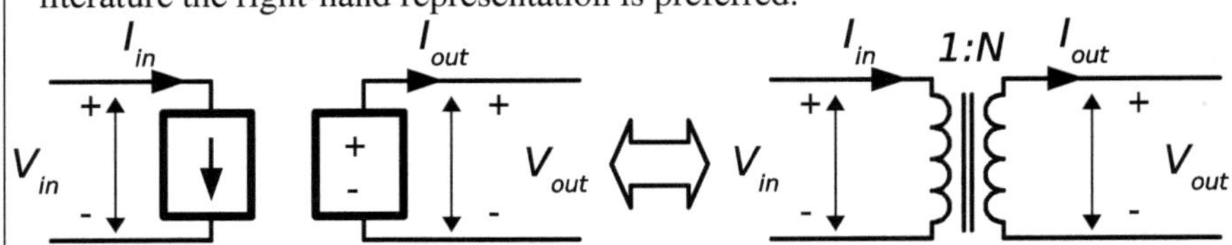

According to the Branch Analysis the output impedance R_{out} has a dual nature. The Slow Switching nature includes the losses associated with switching the capacitors. While the Fast Switching nature includes nothing but the resistive losses of the converter. This dichotomy is based on the dual nature of the branch analysis, performed in Chap. 2. In case the resistive losses are negligible, the converter operates according to the Slow Switching Limit (SSL). In case the resistive losses govern the capacitive converter's behavior, the converter operates according to the Fast Switching Limit (FSL). This literally means that the output impedance in the first case corresponds to the Slow Switching Limit Output Impedance and in the latter case to the Fast Switching Limit Impedance. These both modes of operation correspond to the Full Charging Mode and Partial Charging Mode discussed in Chap. 2.

In the Slow Switching Limit the output impedance R_{SSL} is inversely proportional to the switching frequency f_{sw} of the DC–DC converter and to the total amount of flying capacitance C_{fly}. The proportionality factor K_c depends on the topology and the relative sizing of the flying capacitors.

$$R_{SSL} = \frac{K_c}{C_{fly} f_{sw}} \tag{3.1}$$

The Fast Switching Limit Output Impedance is inversely proportional to the total switch conductance G_{tot}:

$$R_{FSL} = \frac{2K_s}{G_{tot}} \tag{3.2}$$

The proportionality constants K_c and K_s depend on the topology. Calculation of these constants is based on the optimization method presented in Seeman and Sanders (2008). K_c is defined as follows:

$$K_c = \Sigma_i a_{c,i}^2 \tag{3.3}$$

for which a_{ci} are the elements of the charge multiplier vector a_c.

Similarly K_s is defined as follows:

$$K_s = \Sigma_j a_{s,j}^2 \tag{3.4}$$

for which a_{sj} are the elements of the switch vector a_s.

In Fig. 3.2 the switching limit asymptotes of a capacitive DC–DC converter are plotted on a semi-log graph. The horizontal axis represents the switching frequency of the converter, the vertical axis represents the output impedance. The Slow Switching Limit demonstrates a linear, negative slope, behavior with respect to the switching frequency. The Fast Switching Limit is constant since the Fast Switching Limit impedance is independent of the switching frequency.

Both the Slow Switching and the Fast Switching Output impedance can be influenced by changing the parameters of the capacitive converter. For example by increasing the switching frequency, the operation-point of the converter shifts form [a] to [b], or by simultaneously increasing the switching frequency and reducing the amount of flying capacitance, the operation-point of the converter moves from [a] to [c]. Also the Fast Switching Limit impedance is subject to the parameter variations: by increasing the total conductance of the switches the output impedance is decreased and the operation point of the converter shifts form [d] to [e] .

A real capacitive DC–DC converter is subject to both the capacitive and the resistive losses. Thus the output impedance has both a Slow Switching component as a Fast Switching Component. In literature, two approaches are found Seeman and Sanders (2008), the first consider the sum of both components as an approximation of the total output impedance:

$$R_{out} = R_{SSL} + R_{FSL} \tag{3.5}$$

The other approaches found in the square of the quadratic sum a more fitting approximation, shown in Fig. 3.3. In general the latter approximation Seeman and Sanders (2008), Van Breussegem and Steyaert (2011) is preferred:

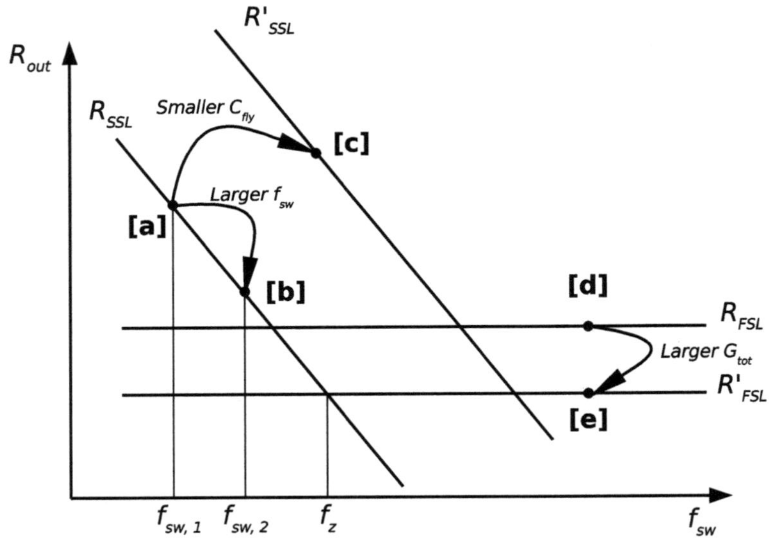

Fig. 3.2 Output impedance limits in function of the switching frequency and subject to different parameter variations

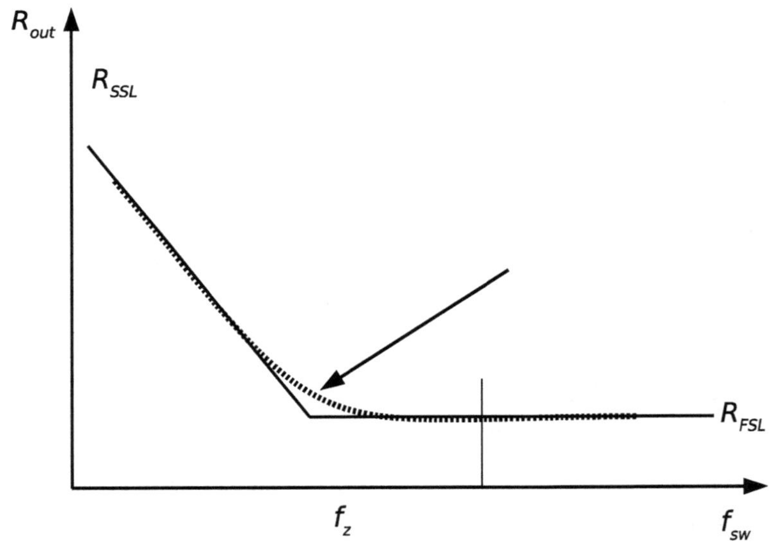

Fig. 3.3 Output impedance limits in function of the switching frequency

$$R_{out} = \sqrt{{R_{SSL}}^2 + {R_{FSL}}^2} \tag{3.6}$$

3.2 Design of Single-Topology Single-Operation-Point Converters

This design approach is called Output Impedance Balancing (OIB). This OIB technique is based on the relationship between the FSL- and the SSL-nature of the converter and the converter performance. The Output Impedance Balancing technique demonstrates how to optimize the ratio of the FSL-SSL-nature towards maximum efficiency. The first part of the following derivation describes the set of generic implementation details and the relationship between the DC–DC converter's requirements and the output impedance. Next the Output Impedance Balance Factor (β) is introduced, to quantify the ratio of the SSL- and FSL-nature. Finally the losses in the DC–DC converter are analyzed and minimized with respect to β. By doing so a closed form optimization is developed for designing a single-topology capacitive converter.

3.2.1 Implementation Parameters

In order to design a real capacitive converter, a first refinement of the Output Impedance Model is required. This additional refinement includes all the relevant implementation details: those which significantly influence the converter performance. In Chap. 1 the primary performance metrics of DC–DC converters were enumerated, the steady-state performance is characterized by:

- Conversion Efficiency
- Maximum Output Power
- Output Power Range
- Operation Points

In order to include the relevant implementation details a closer look at the implementation of a capacitive converter in CMOS is required. The conversion block of a capacitive converter basically exists of a set of switches and a set of capacitors. The analysis in the previous chapters assumed an implementation with ideal switches and ideal capacitors . From this point onwards a number of additional characteristics are taken into account. In the next paragraphs a set of parasitic parameters is added to the existing model.

Capacitors

In general the capacitor implementation is characterized by a series resistance and a parasitic capacitance as is demonstrated in Fig. 3.4a. In a first approximation the parasitic resistance, known as the equivalent series resistance R_{esr}, is neglected. But doing so requires post-layout validation. The parasitic capacitance denotes the

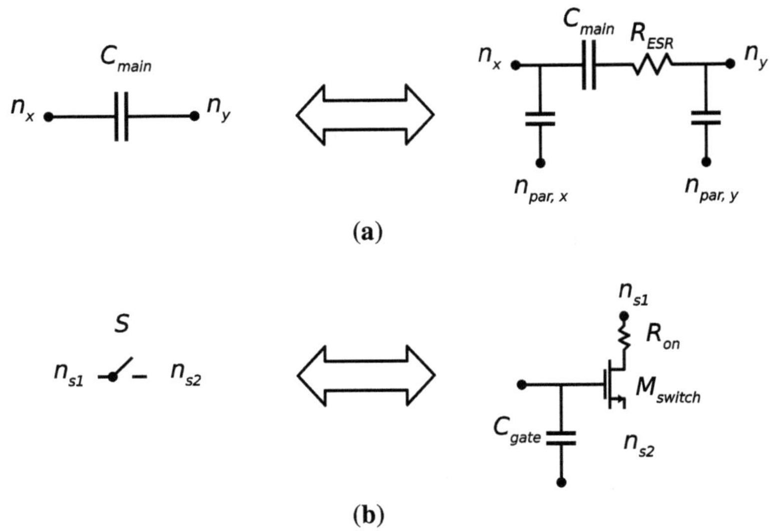

Fig. 3.4 Model refinement

capacitive coupling between the capacitor plates/connections and the surrounding circuits or connections. This parasitic capacitance is modeled as a single capacitor at each node $n_{x/y}$ to an arbitrary other node $n_{par,x/y}$. The main capacitance is a function of the volume occupied by the capacitance and the capacitance density is related to the nature and realization of the capacitor. In CMOS technology the capacitor is constrained to a fixed thickness and the main capacitance is proportional with the capacitor's area. In Chap. 6 the different capacitor types in CMOS are analyzed and mapped on this model.

Switches

For implementation of the switches in a capacitive DC–DC converter CMOS transistors or other solid-state switches are used. The characteristics of these type of switches are discussed in Chap. 6. But until then we take a finite on-resistance into account and an energy penalty to change the switch's position. The on-resistance R_{on} is:

$$R_{on} = \frac{1}{K_N \frac{W}{L}(V_{gs} - V_{th})} \tag{3.7}$$

K_N and V_{th} are technology and device constants, W L are the switch size parameters and V_{gs} is a design parameter. The energy penalty related to changing the state of the switch (on or off) is modeled as the power loss occurring during the charging and discharging of a capacitor. This equivalent capacitor has a capacitance C_{gate} which

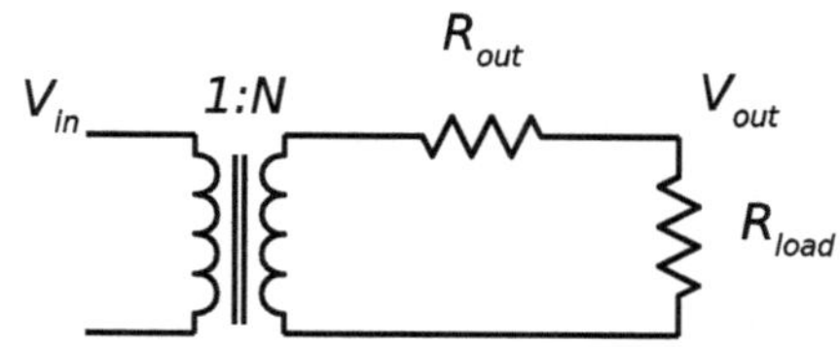

Fig. 3.5 Output impedance model of a capacitive DC–DC converter in case a resistive load is applied

is proportional to the net area of the switch ($W \times L$) and the proportionality constant C_{sq}, is a technology constant:

$$C_{gate} = C_{sq} W L \tag{3.8}$$

It is readily observed that a small parasitic on-resistance results in a large switch (large W since L has a lower boundary). And a large switch leads to a large switch capacitance C_{gate} and inevitably high power loss when operated.

3.2.2 *Output Impedance Requirements*

First the minimum output impedance of the capacitive DC–DC is determined. This minimum required output impedance manifests itself at the maximum load condition for the operation point and the topology under consideration. In Fig. 3.5 the equivalent model of capacitive converter with a resistive load is shown.

Section 3.1 explains that a capacitive DC–DC converter has a non-zero output impedance. Due to this non-zero output impedance, the actual output voltage of the DC–DC converter is different from the unloaded ideal output voltage. The difference between the actual and the ideal output voltage is quantified by γ. Equation (3.9) defines γ as the ratio of the actual output voltage $V_{out,wanted}$ and the ideal (unloaded) output voltage NV_{in}.

$$\gamma = \frac{V_{out,wanted}}{NV_{in}} \tag{3.9}$$

But in Fig. 3.5 it is demonstrated that γ corresponds also to the resistive division between the equivalent load resistance and the output impedance:

$$\gamma = \frac{R_{load}}{R_{load} + R_{out}} \tag{3.10}$$

The output voltage and hence γ is determined by the converters specifications through the relationship with $V_{out,wanted}$. When the required γ is combined with the equivalent load R_{load}, the maximum output impedance value R_{out} can be calculated

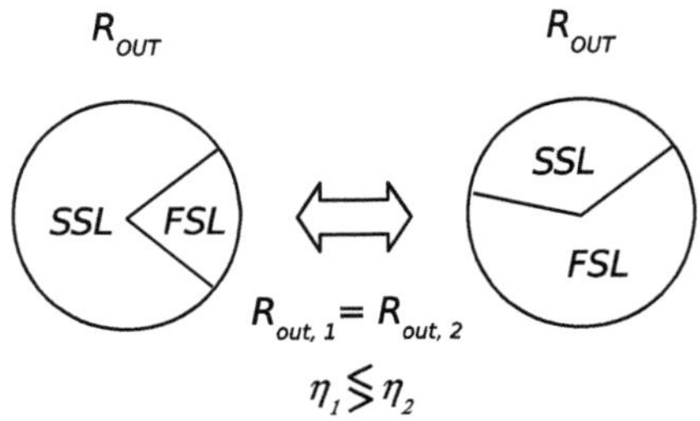

Fig. 3.6 Capacitive converter can be designed with different parameters, resulting in identical output impedance but totally different efficiency

as follows:

$$R_{out,min} = R_{load,max} \frac{1-\gamma}{\gamma} \tag{3.11}$$

3.2.3 Output Impedance Balance

In Sect. 3.1 the dual nature of the output impedance is discussed. And it is demonstrated that combining both natures results in an excellent approximation of the converter's actual output impedance. Clearly a required output impedance can be a achieved by either combining a large SSL component with a small FSL component or vice versa, this is demonstrated in Fig. 3.6. Thus the optimum combination of the SSL and FSL nature has to be determined. The optimum combination is that for which the performance is maximized.

Until now the only relationship between the FSL nature and the SSL nature is that the quadratic sum results in the square of the total output impedance. Hence the Output Impedance Balance Factor β is introduced, to quantify the relationship between both natures. The SSL-nature of the output impedance is proportional to the square root of β:

$$R_{SSL} = \sqrt{\beta} R_{out} \tag{3.12}$$

The FSL-nature of the output impedance is proportional to the square root of β's complement:

$$R_{FSL} = \sqrt{1-\beta} R_{out} \tag{3.13}$$

These definitions fit the approximation of the total output impedance:

$$R_{out} = \sqrt{R_{SSL}^2 + R_{FSL}^2} = \sqrt{1-\beta+\beta} R_{out} \tag{3.14}$$

3.2.4 Parameter Substitution

For a capacitive DC–DC converter the primary characteristics are the total amount of flying capacitors C_{fly}, the switching frequency f_{sw} and the total size of the switches (proportional to the total switch conductance G_{tot}). These characteristics were already introduced in Sect. 3.1. These characteristics can be expressed as a function of the newly introduced Output Impedance Balancing Factor.

The switching frequency f_{sw} is isolated from Eq. (3.1), substituted by formula (3.12) and expressed in function of the topology constant K_c, β and the total output impedance R_{out} in:

$$f_{sw} = \frac{K_c}{\sqrt{\beta} R_{out} C_{fly}} \tag{3.15}$$

The total switch conductance G_{tot} is isolated from Eq. (3.2) and expressed in function of the topology K_s, β and the maximum output impedance R_{out} in:

$$G_{tot} = \frac{2K_s}{\sqrt{1-\beta} R_{out}} \tag{3.16}$$

Since the DC–DC converter's area is dominated by the amount of flying capacitors, C_{fly} is based on the available/acceptable chip area for the design and is assumed to be a constant. The conductance of each individual switch G_j is defined as a fraction $a_{s,j}$ of the total switch conductance G_{tot}

$$G_j = \frac{a_{s,j}}{\Sigma_j a_{s,j}} G_{tot} \tag{3.17}$$

While the individual width of the switches W_j is deducted from the conventional resistance formulation of the CMOS-switch. Eq. (3.18) gives the relationship between switch width W_j, the switch conductance G_j, the minimum feature size of the technology L_{min}, the technology constant K_p, the gate voltage of the switch $V_{gate,j}$ and the threshold voltage of the switch V_{th}.

$$W_j = \frac{G_j L_{min}}{K_p (V_{gate,j} - V_{th})} \tag{3.18}$$

Substitution of Eqs. (3.16), (3.17) and Eq. (3.18) gives Eq. (3.19), which gives the relationship between switch sizing and β.

$$W_j = \frac{\frac{a_{s,j}}{\Sigma_j a_{s,j}} \frac{2K_s}{\sqrt{1-\beta} R_{out}} L_{min}}{K_p (V_{gate,j} - V_{th})} \tag{3.19}$$

As the primary design characteristics are defined as a function of the system requirements, the technology constants and the OIB-factor β. In the next section the DC–DC converter's losses are depicted as a function of β.

3.2.5 Loss Analysis

The converters efficiency is calculated by means of the output power and the power loss during operation. Therefore the primary losses are enumerated. One can distinct two types of losses in a capacitive DC–DC converter: the intrinsic losses and the extrinsic losses. The intrinsic losses appear due to the nature of the DC–DC converter. Because of these losses: the maximum efficiency of the DC–DC converter corresponds to γ. γ is not equal to 100 % due to the non-zero output impedance. The extrinsic losses are the losses introduced by the implementation. They further depreciate the DC–DC converter by introducing an additional decrease in efficiency. But since these extrinsic losses are invoked by the implementation, they can be minimized by an appropriate design approach.

The dominant extrinsic losses are the capacitive losses P_{gate} that originate from charging and discharging the gates of the solid state switches:

$$P_{gate} = \Sigma_j {V_{gate,j}}^2 f_{sw} C_{gate,j} \tag{3.20}$$

In Eq. (3.20) $V_{gate,j}$ is the voltage swing at respectively the gate of MOS switch j. This switch has a total gate capacitance $C_{gate,j}$.

Besides the gate losses also losses due to charging and discharging the parasitic capacitance of the flying capacitors arise, most types of capacitors have a dominant parasitic capacitance.

$$P_{par} = \Sigma_i {V_{nodes,i}}^2 f_{sw} C_{par,i} \tag{3.21}$$

In Eq. (3.21) $V_{nodes,i}$ corresponds to the voltage swing at the nodes connected to the ith flying capacitor's bottom plate.

The power lost due to the finite output impedance in Eq. (3.22) is modeled as a power loss $P_{R_{out}}$ in a resistor due to a DC load current and is the intrinsic loss.

$$P_{R_{out}} = \frac{(NV_{in} - \gamma N V_{in})^2}{R_{out}} \tag{3.22}$$

The efficiency is then formulated as the ratio of the output power and the total input power- including the power losses:

$$\eta = \frac{P_{load,wanted}}{P_{load,wanted} + P_{gate} + P_{par} + P_{R_{out}}} \tag{3.23}$$

Now the information from previous sections is used to reformulate the loss formula's in function of the well known design characteristics. The gate capacitance $C_{gate,j}$ of the CMOS switches equals $C_{sq} W_j L_{min}$ and by substitution of Eqs. (3.17) (3.18) in Eqs. (3.20) (3.21), the gate losses are expressed in function of the system characteristics and the square root of β as well as the square root of β-s complement in:

$$P_{gate} = \frac{K_c}{\sqrt{\beta} R_{out} C_{fly}} \frac{2K_s}{\sqrt{1-\beta} R_{out}} \Sigma_j V_{gate,j}{}^2 C_{sq} \frac{L_{min}}{K_p(V_{gate,j} - V_{th})} \frac{a_{s,j}}{\Sigma_j a_{s,j}} \tag{3.24}$$

The parasitic losses demonstrate a dependency only on the square root of β:

$$P_{par} = \frac{K_c}{\sqrt{\beta} R_{out} C_{fly}} \Sigma_i V_{nodes,i}{}^2 C_{par,i} \tag{3.25}$$

The power loss due to the output impedance is independent from β and is a constant based on the converter's requirements:

$$P_{R_{out}} = \frac{(NV_{in} - \gamma N V_{in})^2}{R_{out}} \tag{3.26}$$

3.2.6 Loss Minimization

Finally the losses of the DC–DC converter are minimized to obtain a DC–DC converter that operates in the most efficient way, given a certain set of requirements and characteristics. Maximizing the efficiency in Eq. (3.23) corresponds to minimizing the losses in:

$$P_{Loss} = P_{gate} + P_{par} + P_{R_{out}} \tag{3.27}$$

The previous equation can be reduced to:

$$P_{Loss} = \frac{K_1}{\sqrt{1-\beta}\sqrt{\beta}} + \frac{K_2}{\sqrt{\beta}} + K_3 \tag{3.28}$$

The technology constants, the DC–DC converter's characteristics and the specifications are lumped into respectively K_{1-3}.

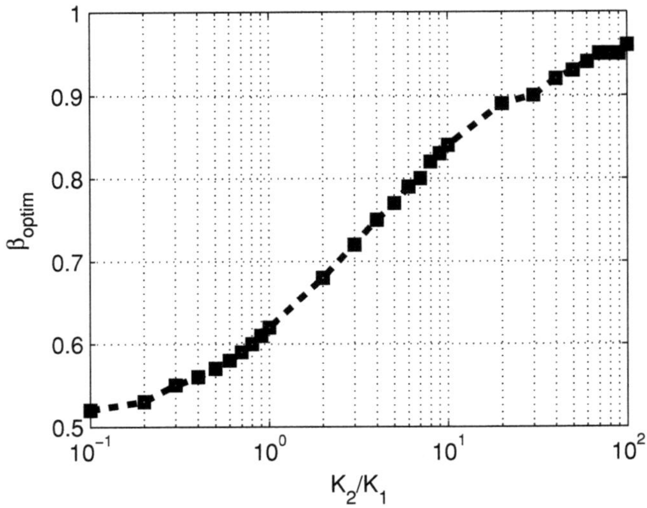

Fig. 3.7 Optimum β as a function of the ratio $\frac{K_2}{K_1}$

K_1 is a constant that corresponds to the gate loss term:

$$K_1 = \frac{K_c}{R_{out}C_{fly}} \frac{2K_s}{R_{out}} * \Sigma_j V_{gate,j}{}^2 C_{sq} \frac{L_{min}}{K_p(V_{gate,j} - V_{th})} \frac{a_{s,j}}{\Sigma_j a_{s,j}} \tag{3.29}$$

K_2 reflects the losses due to the parasitic capacitance in the circuit:

$$K_2 = \frac{K_c}{R_{out}C_{fly}} \Sigma_i V_{nodes,i}{}^2 C_{par,i} \tag{3.30}$$

While K_3 is equal to the losses due to the output impedance:

$$K_3 = \frac{(NV_{in} - \gamma N V_{in})^2}{R_{out}} \tag{3.31}$$

In Eq. (3.28) only β is an independent variable considering fixed requirements, specifications and technology characteristics. Therefore we can minimize the losses by differentiating towards β and solving the characteristic equation.

3.2.7 Analysis

The most important characteristic of Eq. (3.28) is the fact that the constants K_{1-3} are fully determined by the operation-point of the converter and the selected technology. Moreover, differentiation towards β eliminates the constant term and thus also the

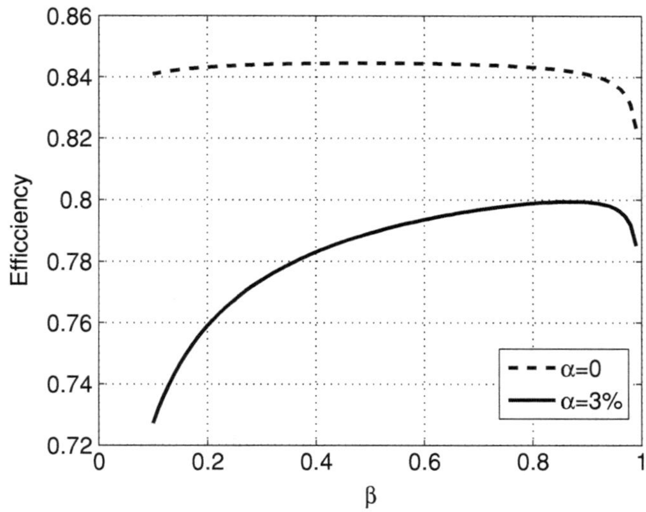

Fig. 3.8 Balancing factor in two different cases. First in case the flying capacitors are not subject to parasitic coupling to the ground ($\alpha = 0\,\%$). Secondly in case that there is significant capacitance between the flying capacitor's plates and DC node in the circuit ($\alpha = 3\,\%$)

parameter K_3. This means that the optimum β only depends on the ratio $\frac{K_2}{K_1}$. In Fig. 3.7 this is calculated for a probable range of the balance ratio. If no parasitic capacitance is present, $K_2 = 0$, the efficiency is maximum for $\beta = 0.5$ irrespective of the converter's requirements or the characteristics. However if parasitic capacitance is present, β is function of the technology characteristics and the specifications. This is demonstrated in Fig. 3.8. First in case the flying capacitors are not subject to parasitic coupling to the ground ($\alpha = 0\,\%$). Next in case there is, there is significant capacitance between the flying capacitor's plates and DC node in the circuit(α=3 %).

3.3 Design of Multi-Topology Converters

The previous section elaborated on a comprehensive design approach, the Output Impedance Balancing technique, for single topology—single-operation-point capacitive DC–DC converters. A large selection of the commercially available discrete-type capacitive converters (Max, 1682), (Max, 16945) and state-of-the-art integrated types Vincent et al. (2009), Somasekhar et al. (2010), Breussegem and Steyaert (2011) fall within this category. The development of evermore power-dense and high-performance capacitive DC–DC converters led to the increasing interest in flexible reconfigurable multi-topology converters, proposed in Chap. 2. But the design of a capacitive DC–DC converter containing multiple topologies introduces three additional constraints. The design should:

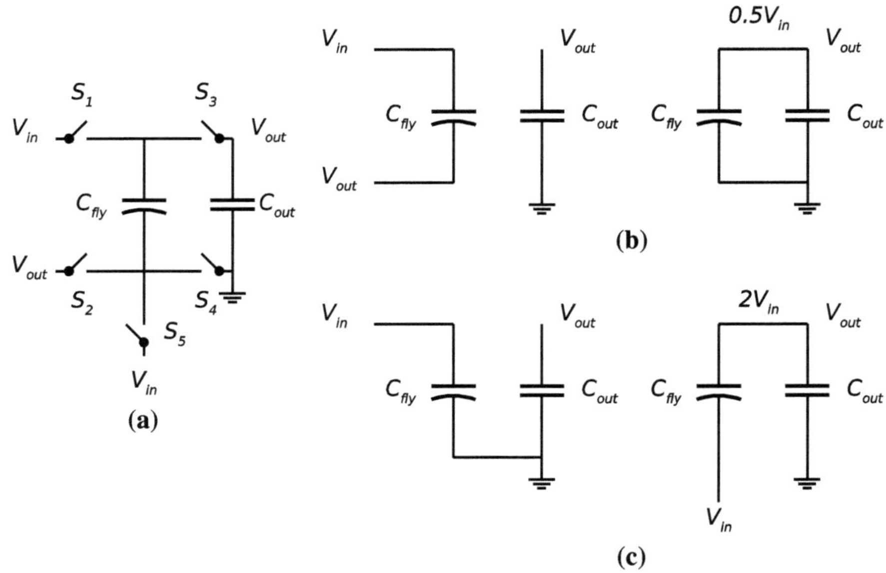

Fig. 3.9 **a** A multi topology capacitive converter **b** The voltage-divider configuration **c** The voltage-doubler configuration

- Include the influence of the additional idle switches
- Meet the requirements for a set of operation points
- Give an optimum solution given a set of user constraints

In order to meet these requirements the converter model described above is extended and a multi-objective algorithm is used for the purpose of optimizing the design.

3.3.1 Model Refinement

The model used in the previous sections describes a single-converter topology and thus takes only the switches and capacitors of that specific topology into account. If a multi-topology converter is observed, the topology consists of operating components and idle components. The operating components belong to the topology addressed for the conversion of interest, while the idle components belong to the topology which addresses another part of the conversion range. The idle components do not participate in the conversion, but since they are physically connected to the operating components, they influence the operating part of the circuit.

In Fig. 3.9 the combined converter including a voltage divider and a voltage doubler is shown. Switches $S_{1-2-3-4}$ are operating in case the voltage divider is used. Switches $S_{1-3-4-5}$ are operating in case the voltage doubler is used. In Fig. 3.10

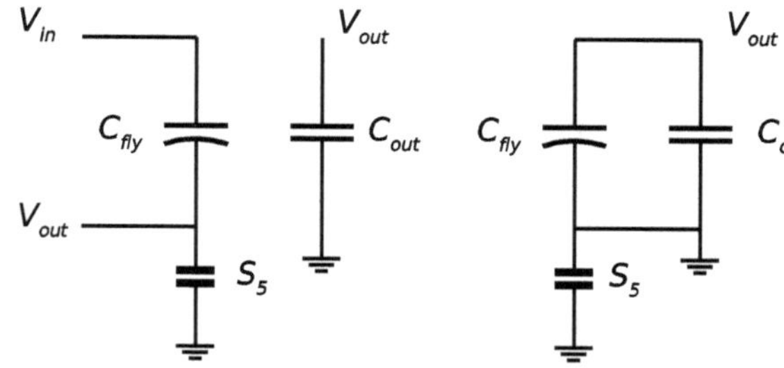

Fig. 3.10 Voltage-divider topology including a lumped parasitic capacitor due to the presence of switch S_5 of the voltage doubler topology

the voltage divider operation is demonstrated, including the influence of the other topology. The idle switch S_5 is modeled as a parasitic capacitor between the flying capacitors node and the ground reference. Although the switch S_5 is not part of the divider topology, he loads the flying capacitor's terminal. Actually the parasitic capacitance is charged by the output capacitor during the first state (left pane of Fig. 3.10) and discharged to ground during the second state (right pane of Fig. 3.10). In this example the switch overhead is 20 % due to the single additional switch.

The parasitic capacitive loading by the idle switches is taken into account by adding an additional loss term to the loss set defined in the Sect. 3.2 of this chapter. The energy loss equals the energy that is transferred by the parasitic capacitance each cycle. The charge involved is not transferred towards the load but is dissipated from the load towards the ground. In fact the efficiency of the converter providing this charge to the load should be taken into account, but since we strive for a high efficiency this effect can be neglected. Therefore the power loss due to the parasitic capacitance is approximated by:

$$P_{par,switches} = \Sigma_j V_{node,j}{}^2 f_{sw} C_{par,switch,j} \tag{3.32}$$

By substituting Eq. (3.15) and considering that the parasitic capacitance introduced by the idle switches is proportional with the switch size, the following expression of these losses is derived:

$$\begin{aligned} P_{par,switches} &= \frac{K_c}{\sqrt{\beta} R_{out} C_{fly}} \frac{2K_s}{\sqrt{1-\beta} R_{out}} \cdot \\ \Sigma_i V_{nodes,i}{}^2 (\Sigma_k C_{sq} & \frac{L_{min}{}^2}{K_p (V_{gate,k} - V_{th})} \frac{a_{s,k}}{\Sigma_j a_{s,j}}) \end{aligned} \tag{3.33}$$

This loss term is of the same shape as the loss term associated with the switching losses: Inversely proportional to both the square root of β and the square root of β-s complement. This gives again a loss equation of the form demonstrated in Sect. 3.2 Eq. (3.28):

$$\begin{aligned} P_{Loss} &= (P_{gate} + P_{par,switches}) + P_{par} + P_{R_{out}} \\ P_{Loss} &= \frac{K_1}{\sqrt{1-\beta}\sqrt{\beta}} + \frac{K_2}{\sqrt{\beta}} + K_3 \end{aligned} \tag{3.34}$$

The lumped constant K_1 not only reflects the gate switching losses but also the switch-related parasitic terms. Again the efficiency of a single topology can be maximized by minimizing with respect to β for a single topology including the capacitive parasitics of the idle topology.

3.3.2 Optimization Space

The multi-topology capacitive converter requires a different design approach than the single-topology and single-operation-point capacitive converter. First instead of defining a single critical operation point a set of operation points R is defined :

$$R_x \in R \tag{3.35}$$

In the simplest case this set includes a combination R_x of each required input-output voltage including the load current for each input-output voltage combination:

$$R_x = [V_{in,x} \quad V_{out,x} \quad I_{load,x}] \tag{3.36}$$

It is impossible to dimension and optimize each topology separately since part of the switches operate in multiple configurations, this would invoke conflicting sizing instructions. Moreover the sizing of one topology influences the parasitics in the complementary topologies.

This type of problems can be addressed by multi-objective optimization approaches. In this problem a set of system parameters is proposed in order to obtain a system performance conform the a priori defined requirements and maximizing the overall system performance. The set of system parameters includes the switch size and potentially also the capacitor size. The set of requirements is defined by R and the performance is expressed by means of the system efficiency. In a typical case the number of parameters is larger than ten and according to the Hughes Effect, the search space becomes multi-dimensional and the complexity of the space increases at a high rate. A number of advanced algorithms exist to solve these kind of problems: for example the genetic algorithm Jun Zhang et al. (2001) or the ant colony approaches Jun Zhang et al. (2009). But all of these techniques involve exploration of the large multi-dimensional search space.

> In applied mathematics,the curse of dimensionality also known as the Hughes effect or Hughes phenomenon (named after Gordon F. Hughes) refers to the problem caused by the exponential increase in volume associated with adding extra dimensions to a mathematical space. (Wik)

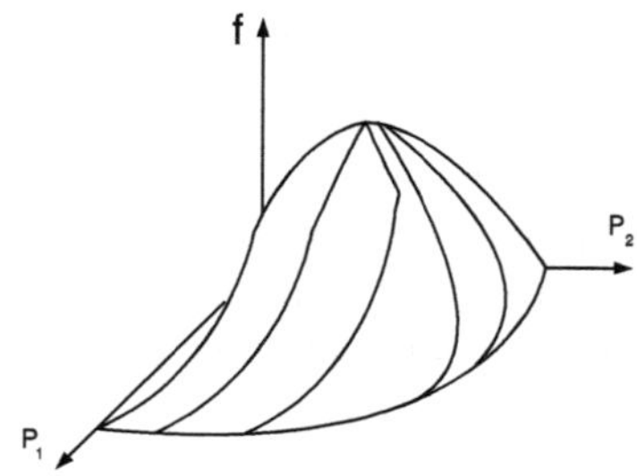

Fig. 3.11 Generic representation of an optimization problem including 2 parameters P_1 P_2 and the resulting objective function f which must be maximized

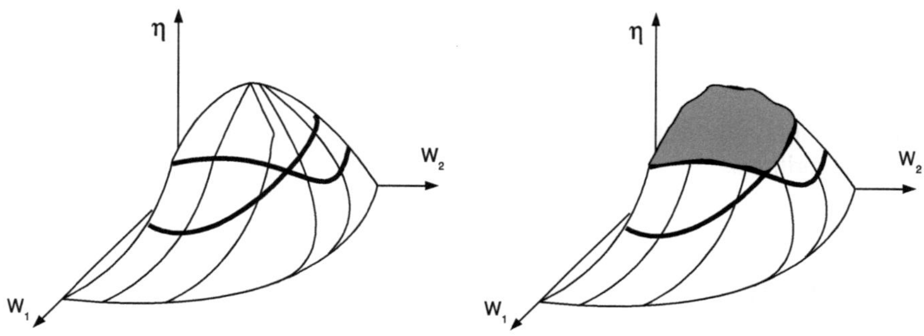

Fig. 3.12 Generic representation of a converter optimization problem including 2 parameters W_1 W_2 and the resulting objective function: efficiency η which must be maximized. **a** Two operation-points determine maximum limits on the potential optimization space and constraint thus this optimization space. **b** The optimization subspace

Given that not only the objective function, reflecting the converter's performance, must be optimized but both requirements must be met for every operation point, an additional constraint is defined: the existence constraint. This constraint can be used to confine the search space by eliminating parts of the space which are ruled out because of the existence constraint. This is an advantage in case this constraint is defined by means of little computation (Fig. 3.11).

Actually a constraint for the switch sizing can be defined by means of the output impedance-balancing method. For each topology separately the range of potential switch sizes is given for the range corresponding to $\beta = 0$ and $\beta = 1$. So by solving for each operation point and corresponding topology the minimum and maximum size of the switches the optimization space is constraint to a subspace. The switch size corresponding to $\beta = 0$ corresponds to the minimum switch size while $\beta = 1$ corresponds to the maximum switch size (Fig. 3.12).

3.3.3 Multi-Objective Optimization

For optimization of the converter a search algorithm is operated on the converter's objective function. The objective function of this converter corresponds to a weighted average of the converter efficiency over all requirement sets R_x. In case of a uniform probability density function the following objective function is required:

$$f(R_x) = \frac{1}{p}\Sigma_{x=1}^{p}\eta(R_x) \tag{3.37}$$

$$\eta(R_x) = \frac{P_{out,R_x}}{P_{out,R_x} + P_{loss,R_x}} \tag{3.38}$$

The optimum for the objective function can be found by means of different types of search algorithms.

3.4 Accuracy Improvement

It has been demonstrated in the previous sections that the design of capacitive DC–DC converters requires accurate modeling and analysis. But the conventional Output Impedance Model, as presented and used in the previous sections, fails to include the influence of the output buffer capacitor size on the performance of the capacitive DC–DC converter. This model is conceived based on the assumption that capacitive DC–DC converters have an infinitely large output buffer capacitance. To some extent this approximation holds for a lot of the discrete-type converters and part of the fully integrated converters. But for converters with a relatively [1] small output capacitor, significant deviations from the Output Impedance Model are observed. The main reason for fully integrated converters to compromise on output capacitor size, is the high cost of chip area. Hence the buffer capacitance size is decreased and larger levels of output noise are accepted. Section 3.4.1 gives a short overview of the conventional model as it is developed in Seeman and Sanders (2008), Sect. 3.4.2 first gives a physical explanation for the inability of the conventional model to take the output capacitor size into account and next it introduces a mathematical approach of the problem. Finally this section proposes a modified model. This model is applied to the series-parallel type of down-converters in Sect. 3.4.3, verified against *Spice* simulations in Sect. 7.1.4 and validated based on measurements in Sect. 3.4.4.

[1] Relative with respect to the size of the flying capacitors.

3.4.1 Conventional Model

In Seeman and Sanders (2008), the output impedance in the SSL is determined to be:

$$R_{SSL} = \frac{V_{out}}{I_{out}} = \frac{V_{out}}{q_{out} f_{sw}} = \Sigma_i \frac{q_i}{q_{out}^2} \frac{\Delta v_i}{f_{sw}} \tag{3.39}$$

In Eq. 3.39 f_{sw} represents the converters' switching frequency and q_{out} q_i respectively the total charge transferred to the load and the charge transferred by a single capacitor C_i in a topology with n charge transferring capacitors (also known as the flying capacitors). Δv_i is the change in voltage over the i-th flying capacitor. This change in voltage is induced by charge redistribution after connecting capacitor terminals that have different voltage potentials.

If the assumption is made that an infinitely large output buffer capacitor is present the change in voltage Δv_i due to changing phases is only function of the capacitor size C_i and the charge transferred by the capacitor C_i. Thus Eq. 3.16 can be substituted by:

$$\Delta v_i = \frac{q_i}{C_i} \tag{3.40}$$

This equation can be optimized and simplified towards:

$$R_{SSL} = \Sigma_i \frac{a_{ci}^2}{C_{tot} f_{sw}} \tag{3.41}$$

The latter is demonstrated in Seeman and Sanders (2008). C_{tot} represents the total amount of flying capacitance and a_{ci} is the charge transfer vector element corresponding to the ith flying capacitor. In fact $a_{ci} = \frac{q_i}{q_{out}}$.

3.4.2 Modified Model

Physical Approach

The conventional model, revisited in the previous paragraph, is based upon the assumption that an infinitely large output capacitor is present and therefore the output voltage is constant. In practice, the output capacitor is often in the same order of magnitude as the flying capacitors. This gives rise to deviations in output impedance between the measurements and the conventional model: the conventional model gives an overestimate of the output impedance. Which corresponds to an underestimate of the potential converter performance. In order to explain these deviations between the conventional model and the measurements, this paragraph goes deeper into the physical phenomena that appear during the converter's operation. First in case of an infinite output capacitor, next in case of a finite output capacitor.

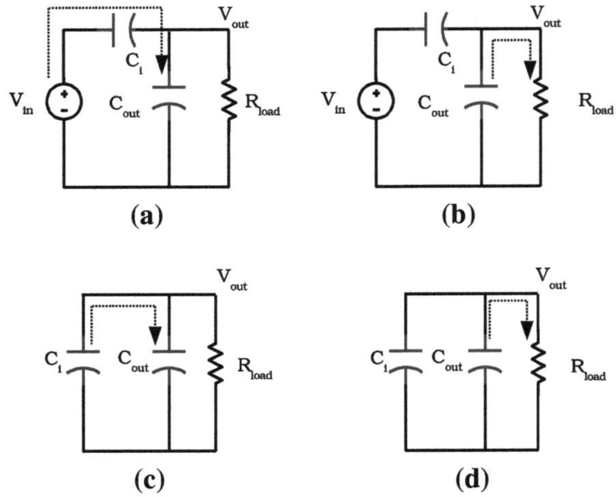

Fig. 3.13 **a** Charge redistribution phase 1 **b** Charge pumping phase 1 **c** Charge redistribution phase 2 **d** Charge pumping phase 2

Infinite Output Capacitance

Two simultaneous phenomena take place during each switching phase: charge redistribution and charge pumping. This is demonstrated for a single flying capacitor converter in Fig. 3.13.

Charge redistribution: Charge is transferred from the flying capacitors to the output capacitor (Fig. 3.13a, c). This is induced by the potential difference of two nodes connected after the reconfiguration of the converter. The charge redistribution associated with this reconfiguration is lossy. By connecting two capacitors a part of the energy on the capacitors is dissipated in the circuit. This will appear as a non-zero output impedance even if ideal switches and ideal capacitors are assumed. The charge redistribution takes place in an impulse-like fashion. This current impulse is damped by the output capacitor and the charge is stored on the output capacitor. Since the output capacitor is infinitely large no variation in output voltage is observed.

Charge pumping: This phenomenon is associated with the transfer of charge from the converter to the load (Fig. 3.13b, d). In fact this comes in the ideal case down to the charge transfer from the output capacitor to the load. Since the output capacitor is nothing but a charge reservoir, all the charge that is transferred from this reservoir was initially transferred from the flying capacitor to the output capacitor.

Charge redistribution transfers charge from the flying capacitor to the output buffer capacitor C_{out}, charge pumping will transfer the charge from the buffer to the load. In case that an infinite output capacitor is used: all charge that is transferred to the load is transferred by a lossy mechanism. The latter gives raise to a non-zero output impedance.

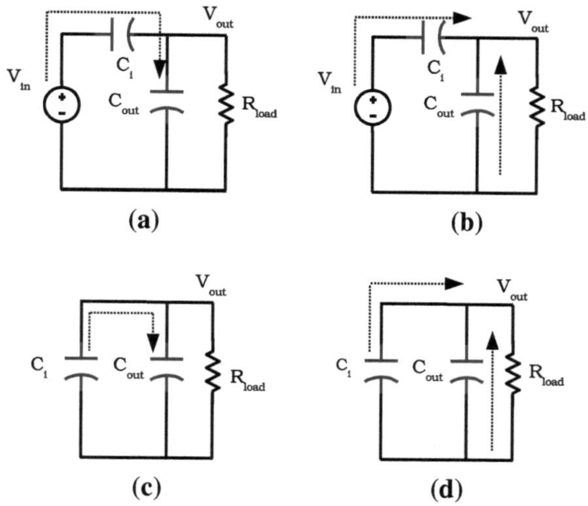

Fig. 3.14 **a** Charge redistribution phase1 **b** Charge pumping phase1 **c** Charge redistribution phase2 **d** Charge pumping phase2

Finite Output Capacitance

The charge redistribution phenomenon that appears when a finite output capacitor is present is identical as in case an infinite output capacitor is present (Fig. 3.14a, c).

In this case charge pumping is different. Since the ouput buffer capacitor has a finite capacitance, the output voltage drops during each phase due to discharging this output buffer capacitor. This change in output voltage induces a change in voltage over the flying capacitor and thus charge transfer from the flying capacitor to the load. Not only the output buffer capacitor will transfer charge to the load, as well will the flying capacitor (Fig. 3.14b, d).

This charge transfer that originates from the flying capacitor corresponds to discharging capacitors by means of a resistive load. This charge transfer is lossless.

This implies that in case a finite output capacitor is used, part of the charge delivered to the load is transferred by means of a lossy mechanism and part of the charge by means of a lossless mechanism. This explains the deviation of the conventional model from the observations in simulations and measurements. In the next paragraph this is quantified and put into an modified/improved Output Impedance Model.

Mathematical Approach

The output impedance is introduced by means of the change in voltage (Δv_i) on the flying capacitors after reconfiguration. According to the conventional model (Eq. 3.39) this Δv_i can be calculated based on the size of the capacitor and the amount of charge transferred to the load by the capacitor. But from the finite output capacitor point of view Δv_i is associated only with the lossy charge transfer thus is induced by the charge redistribution. In order to define the actual Δv_i one has to

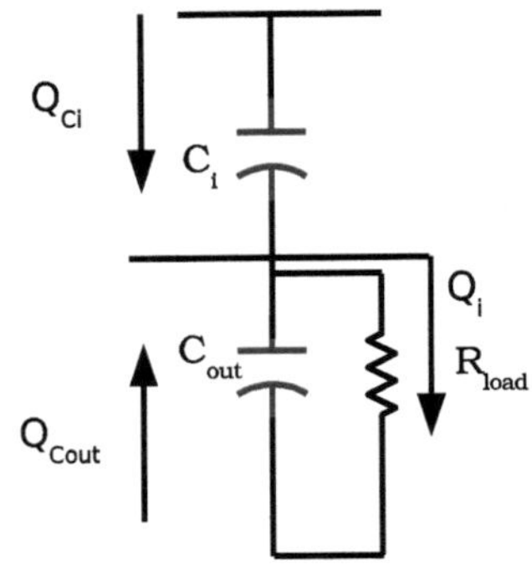

Fig. 3.15 A single flying capacitor voltage divider

quantify the amount of charge that is transferred via the lossy mechanism. The latter corresponds to the charge that originates from the output capacitor C_{out} during the charge pumping. In the next paragraph the ratio between the charge involved in the lossy transfer and the total charge that is delivered to the load is determined.

In Fig. 3.15 a single flying capacitor capacitive DC–DC converter is depicted. During the charge pumping, charge is transferred from as well the flying capacitor q_{C_i} as from the output capacitor q_{Cout} to the load. The total amount of charge (Eq. 3.42) that is delivered to the load is q_i.

$$q_i = q_{Cout} + q_{C_i} \tag{3.42}$$

$$q_i = \Delta r_i Cout + \Delta r_i C_i \tag{3.43}$$

We are looking for the fraction of the charge that is transferred by means of the lossy mechanism thus: $\frac{q_{Cout}}{q_i}$ From Eq. 3.43 it is clear that the change in charge on the capacitors is only function of the size of the capacitors (C_{out} C_i) and the variation of voltage across the capacitor C_i Δr_i. Therefore $\frac{q_{Cout}}{q_i}$ equals:

$$\frac{q_{Cout}}{q_i} = \frac{\Delta r_i Cout}{\Delta r_i Cout + \Delta r_i C_i} \tag{3.44}$$

$$\frac{q_{Cout}}{q_i} = \frac{Cout}{Cout + C_i} \tag{3.45}$$

By calculating the ratio between the charge from C_{out} and the charge that is delivered to the load (Eq. 3.44), it is shown that this ratio is only function of the capacitor sizes (Eq. 3.45). The ripple Δr_i itself is a function of the other system parameters, such as the switching frequency and the load, but is cancelled out. Based upon 3.45 a new formula for Δv_i is proposed:

$$\Delta v_i = \frac{q_i}{C_i} \frac{C_{out}}{C_{out} + C_i} \tag{3.46}$$

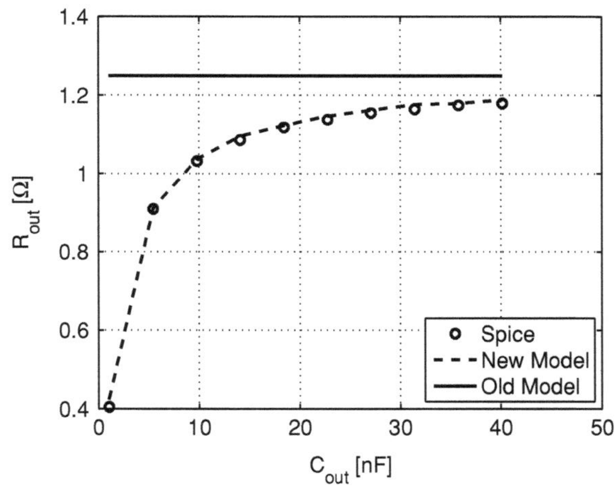

Fig. 3.16 Output impedance of a series parallel 1/2 capacitive DC–DC Converter as a function of the output capacitor size

This change in voltage is function of the charge that is transferred by means of the lossy mechanism. So that a modified Output Impedance Model is proposed:

$$R_{SSL,modified} = \Sigma_i (\frac{q_i}{q_{out}})^2 \frac{C_{out}}{C_{out} + C_i} \frac{1}{C_i f_{sw}} \tag{3.47}$$

$$R_{SSL,modified} = \Sigma_i \frac{C_{out}}{C_{out} + C_i} \frac{a_{ci}^2}{C_i f_{sw}} \tag{3.48}$$

3.4.3 Cases

In this paragraph this improved Output Impedance Model is applied on as well series-parallel topologies as on fractional topologies.

Series-Parallel

In Fig. 3.16 the output impedance of a series-parallel converter with iVCR $\frac{1}{2}$ is plotted in function of C_{out}. The converter has one flying capacitor with a capacitance of 2 nF and a switching frequency of 100 MHz. The C_{out} is swept from 1 nF to 40 nF. The continuous line represents the output impedance value as calculated by means of Makowski's Model and Seemans' optimization Seeman and Sanders (2008): the old model. For small values of C_{out} one can observe a deviation between this model and Spice simulations of up to 200 %. The new Model takes the Output buffer size into account and fit the simulations perfectly.

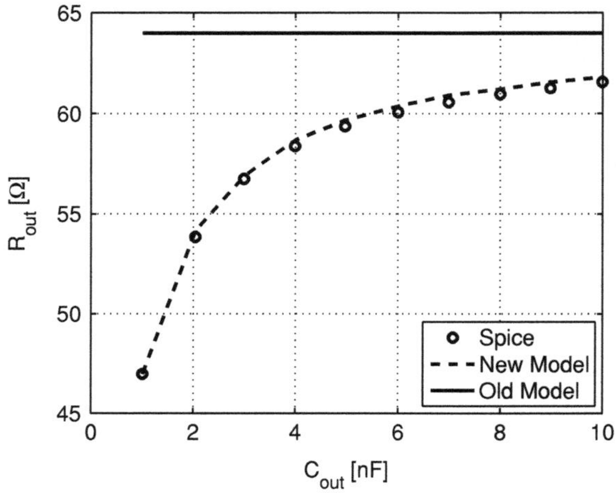

Fig. 3.17 Output impedance of a 4/5 capacitive DC–DC converter as a function of the output capacitor size

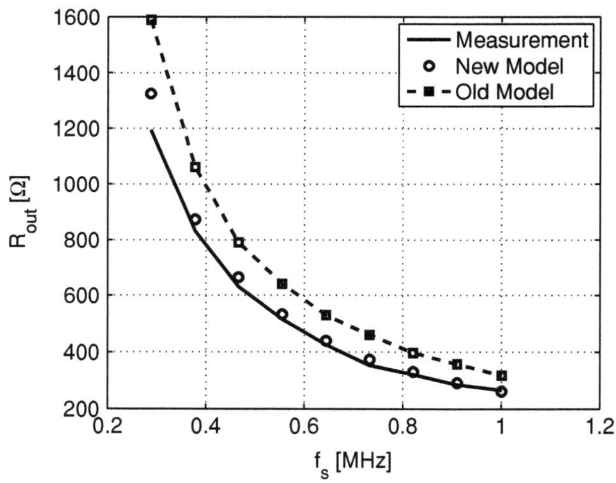

Fig. 3.18 Measurements of the 2/3 converter's output impedance as a function of the switching frequency

Makowski

In Fig. 3.17 the output impedance of a Makowski converter with iVCR of $\frac{4}{5}$ is plotted in function of C_{out}. The converter has three flying capacitors, C_1 C_2 C_3, with respectively a capacitance of 0.25, 0.25 and 0.5 nF. The switching frequency is 10 MHz. The C_{out} is swept from 1 nF to 10 nF. The continuous line in Fig. 3.17 represents the output impedance value as calculated by means of Makowski's Model (1995), and

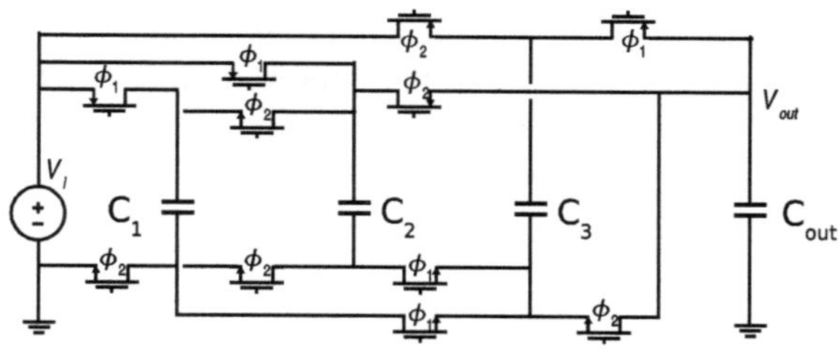

Fig. 3.19 Schematic representation of a capacitive 2/3 converter topology

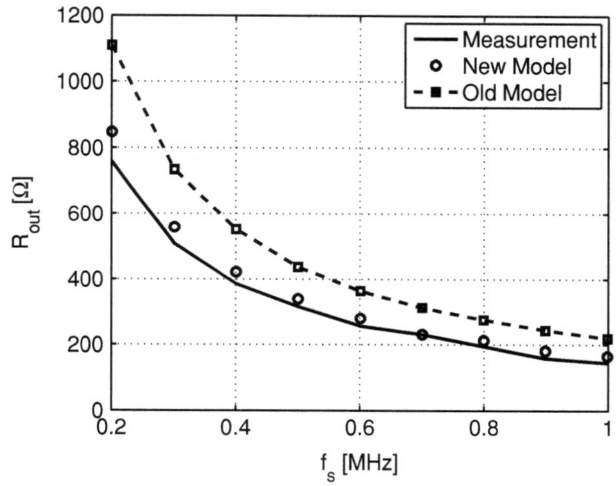

Fig. 3.20 Measurements of the 4/5 converter's output impedance as a function of the switching frequency

Seemans' optimization Seeman and Sanders (2008): the old model. For small values of C_{out} one can observe a deviation between this model and Spice simulations. The new model fits the simulations perfectly.

3.4.4 Measurements

For sake of validation measurements have been performed on two switched-capacitor structures in 90 nm CMOS: a $\frac{2}{3}$-ratio converter and a $\frac{4}{5}$-ratio converter. The results are shown in Fig. 3.19 and in Fig. 3.20. They both have a total flying capacitance of 2 nF and an output buffer capacitance of 3.2 nF. The switching frequency was varied externally in order to retrieve the output impedance in function of the switching frequency. The 4/5-converter has a topology as presented in Sect. 3.4.3 and the 2/3-topology is shown in Fig. 3.18.

The model based on Seeman and Sanders (2008) is marked with the squares, the measurements by the solid line and the improved model by the round marks. Measurements show that in both cases the improved model shows a model accuracy improvement up to 30 %. These measurements validate the improvements made to the conventional model.

3.5 Conclusion

This section presents a closed-form design plan for the capacitive DC–DC converter. Alternative design approaches Su et al. (2010) focus on the optimization of the efficiency while the presented approach highlights the relationship between the optimization parameters and the DC–DC converter's output impedance. The presented optimization has two advantages. First it coincides with the state-of-the-art modeling techniques Makowski and Maksimovic (1995), Seeman and Sanders (2008) and secondly it is not a topology-specific approach, it can easily be applied to other topologies. Next a complementary approach is presented to design multi-topology capacitive DC–DC converters. The model of a single-topology converter is extended and the Output Impedance Balancing method is used to reduce the size of the search space. By doing so the computational effort of the search algorithm is reduced. Finally an accuracy improvement of the Output Impedance Model is discussed. This improvement denotes the inclusion of the output capacitor size in the model.

Chapter 4
Noise Reduction by Multi-Phase Interleaving and Fragmentation

Capacitive DC–DC converters are switched-mode power supplies and thus inevitably generate switching noise. But a fully integrated DC–DC converter is also part of a SoC. Therefore the noise influences the other building blocks of the SoC. It is shown in this chapter that noise mitigation in fully integrated DC–DC converters has a certain cost, related to the chip area that is required to achieve this. The resilience of the other building blocks with respect to the noise generated by the DC–DC converter determines the acceptable level of noise and as a consequence the converter's cost.

In Sect. 4.1 the occurrence of the noise is discussed from the SoC perspective. Next the noise generation in a capacitive DC–DC converter is analyzed in Sect. 4.2. Section 4.3 discusses the effect of noise on a SoC's performance. Finally Sect. 4.4 introduces multi-phase interleaving by fragmentation as a new and effective noise mitigation technique.

4.1 Noise in Systems on Chip

A SoC is a combination of functional blocks monolithically integrated on a single substrate. The fully integrated DC–DC converter is one of them. In Fig. 4.1 a schematic representation is given of a SoC. This SoC includes a single capacitive DC–DC converter and a signal-processing block. This signal-processing block requires two supply voltages: the input voltage V_{in} and a voltage that is generated by the DC–DC converter from the input voltage. In the same figure the most important supply interconnections are drawn. First there are the supply connections between the environment and the chip. These connections are made by means of bond wires. The bond wires are modeled by means of a series inductance $L_{bondwire}$ and a series resistance R_{esr1}, proportional to the bond wire length. The parallel parasitic capacitance is a function of the bond-pad area and bond-pad type. This parallel capacitance is in many cases outnumbered by the intentionally placed input decoupling capacitor $C_{on-chip1}$. The input decoupling capacitor is required to decrease

T. Van Breussegem and M. Steyaert, *CMOS Integrated Capacitive DC–DC Converters*, Analog Circuits and Signal Processing, DOI: 10.1007/978-1-4614-4280-6_4,

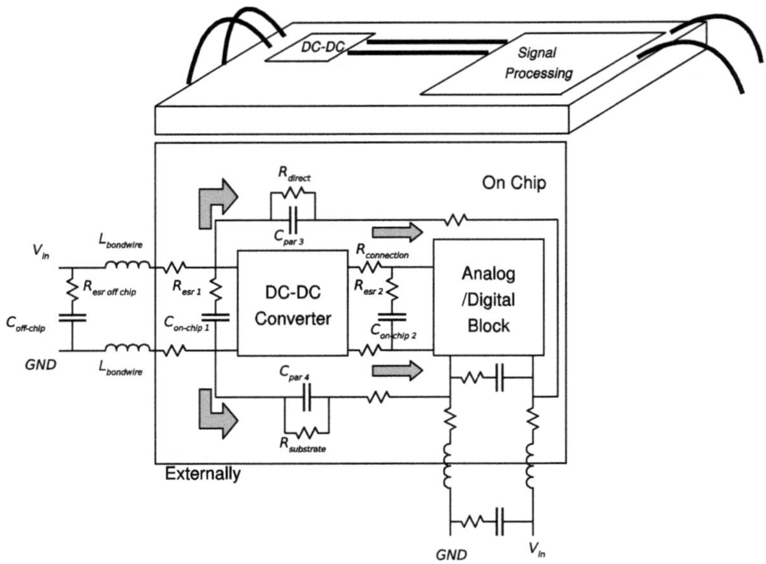

Fig. 4.1 Simplified model of the supply interactions present in a SoC. The *gray arrows* indicate potential noise propagation induced by the pulsed current pattern of capacitive DC–DC converters

the AC impedance of the external supply. The connection between the DC–DC converter and the remaining sub-blocks is modeled by a series resistance $R_{connection}$ and a parallel capacitance $C_{on-chip2}$ with a parasitic resistance R_{esr2}, the inductance is neglected in these relatively short connections. Again, in many cases additional local decoupling is added which over stems the parasitic capacitance. For sake of simplicity the input decoupling of the loading block is absorbed as part of the capacitive DC–DC converter's output capacitance. Parasitic coupling between both supply rails, whether this is direct coupling because of common bond-wire connections R_{direct} or by means of indirect coupling through the substrate $R_{substrate}$, introduces complex interactions between the DC–DC converter and the other system blocks. It is demonstrated in Fig. 4.1 how noise from the DC–DC converter propagates from both the input and the output to the other block in the SoC (Badaroglu et al. 2004).

It is demonstrated in Sect. 2.1 that capacitive DC–DC converters operate by means of pulse-like current patterns. The high-frequency content of the converter's input current in combination with the high inductance of the bond-wires results in high levels of voltage noise (Alon and Horowitz 2008).[1] This effect can be reduced by placing a large decoupling capacitor. The decoupling capacitance, both at the input and the output of the DC–DC converter, increases the area of the converter and thus also the cost of the solution. But since it is better to prevent than to cure,

[1] The voltage across the bond-wire is proportional with the time derivative of the current through the bond-wire: $V_L = \frac{LdI}{dt}$.

especially when the cure has a high cost, the following sections will first analyze the switching noise in capacitive DC–DC converters and next propose a number of intrinsic techniques to reduce the noise coming from the capacitive DC–DC converter.

4.2 Noise Characteristics

In order to model and calculate the switching noise (ripple) in capacitive DC–DC converters, two approaches are followed. The first approach ignores the resistance in the circuit and coincides with the SSL approximation of a converter. This approach gives us a straightforward description, that is easily derived for any converter. The second approach, the FSL approximation, takes the resistance in the circuit into account, but requires more approximations and is less straightforward to be derived. The SSL approach is therefore preferred, especially since the SSL approximation results in the worst-case approximation. Next to the real switching noise, there is also noise which is induced by the imperfections in the components and implementation of the DC–DC converters.

4.2.1 Noise in the Slow Switching Limit

The SSL approximation implies that the charge transfer takes place by means of a current pulse at the instant of commutation, not hindered by the parasitic resistance in the circuit, as is shown in Fig. 4.2. Hence the ripple Δv_{SSL} behaves as a voltage variation of a capacitor connected to a resistive load during a time interval $\frac{T}{\zeta}$. The current, which is drawn from the capacitor, corresponds to the load current of the converter $I_{load}(t) = \frac{V_{out}(t)}{R_{load}}$ and the capacitor corresponds to the total capacitance C^* connected to the output node. In general:

$$\Delta v_{SSL} = V_0 - V_0 e^{\frac{-T}{\zeta R_{load} C^*}} \tag{4.1}$$

In this equation V_0 corresponds to the peak value of the output voltage after commutation. In case that the exponent is relatively small, the equation is linearized by means of a Taylor expansion around the initial voltage V_0 at $t = 0$ and I_{load} corresponds to $\frac{V_0}{R_{load}}$. The ripple can be calculated as follows:

$$\Delta v_{SSL} = \frac{I_{load} T}{\zeta C^*} \tag{4.2}$$

The previous approximation is also valid in case a constant-current load is applied to the capacitive converter. The constant load approximation gives an upper boundary (worst-case value) for the maximum ripple achieved by a resistive load.

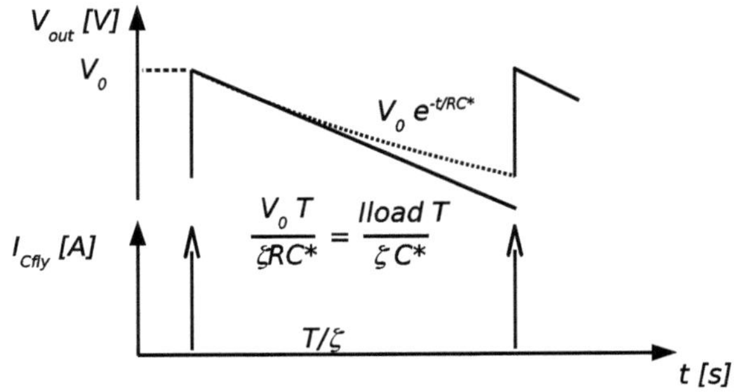

Fig. 4.2 Approximations of the output voltage pattern, assuming SSL prerequisites. The *dashed line* represents a purely resistive load. The *continuous line* represents the constant current load case

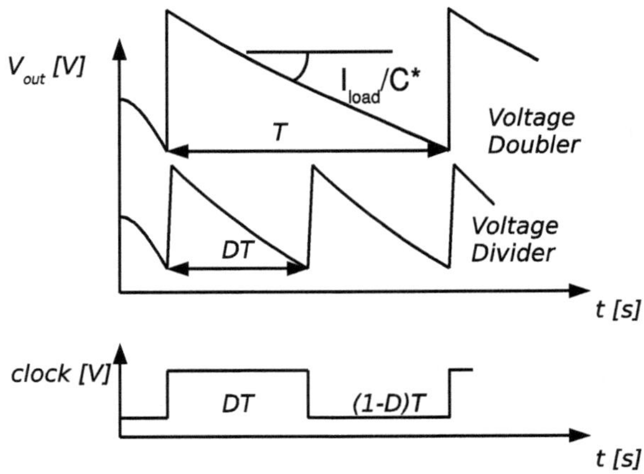

Fig. 4.3 Differentiation between voltage doubler en voltage divider topologies. The voltage divider has two commutation instants during the switching period. The voltage doubler has only one

In Fig. 4.3 the ripple at the output of respectively a voltage doubler and voltage divider is shown. The first critical difference is that the voltage divider initiates a charge transfer at two distinct instants during the switching-period. While the voltage doubler does this at a single point. In case of the voltage doubler, the flying capacitor is connected only during a fraction D of the switching period T and in case of the voltage divider the flying capacitor is connected to the output capacitor at all times-except for the short non-overlap period. Therefore a voltage doubler topology has a $\zeta = 1$ and a voltage divider topology has $\zeta = 2$ (taken $D = 0.5$), the latter resulting in a smaller ripple given the same parameters.

In literature only the impact of the capacitance of the output buffer capacitor C_{out} is taken into account to determine C^* (Sai Kit Lau et al. 2007). While this is a good approximation for converters with a relatively large output buffer capacitor compared to the amount of flying capacitors, the latter does not hold for many of the fully-integrated capacitive converters. For these converters the flying capacitance that is connected to the output will additionally damp the output voltage. In a two-state converter the ripple is determined by the state for which the least amount of flying capacitance is connected to the output node.

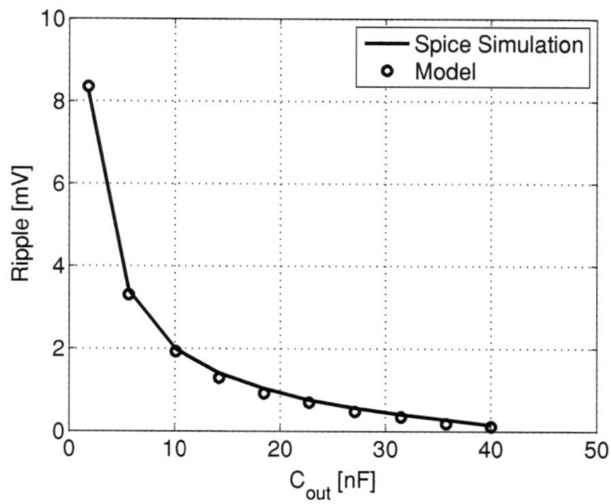

Fig. 4.4 Ripple simulation results of a series-parallel 1/2 converter: The converter has one flying capacitor with a capacitance of 2 nF and a switching frequency of 100 MHz

This introduces an additional implementation-dependent parameter κ_T which represents the ratio of the minimum output connected and the total amount of flying capacitance. For a $1/2$- converter κ_T is equal to 1. C_{fly} is connected to the output terminal during both states. For a $4/5$- converter κ_T is equal to 0.25 for the first phase and 0.625 for the second phase. The resulting-minimum-κ_T is thus 0.25. It is clear that converters have an asymmetric ripple and the weakest damped state determines the worst case ripple in SSL. If the approximation of $C^* = (C_{out} + \kappa_T C_{fly})$ is substituted in Eq. 4.2 and the switching period is substituted by the inverse switching frequency, the following ripple approximation is obtained:

$$\Delta v_{SSL} = \frac{I_{load}}{\zeta f_{sw}(C_{out} + \kappa_T C_{fly})} \tag{4.3}$$

This approach is verified by simulation for the $1/2$-topology and the $4/5$-topology that are described in Chap. 2. In Fig. 4.4 the simulation results of a series-parallel 1/2 converter are shown. The converter has one flying capacitor with a capacitance of 2 nF and a switching frequency of 100 MHz. In Fig. 4.5: the simulation results of a fractional 4/5 converter are depicted. This converter has three flying capacitors, C_1 C_2 C_3, with respectively a capacitance of 0.25 0.25 and 0.5 nF. The nominal switching frequency is 10 MHz. For both cases an excellent match between the SSL approximation and the Spice simulations is obtained. This shows that the model of the output voltage ripple indeed holds in SSL.

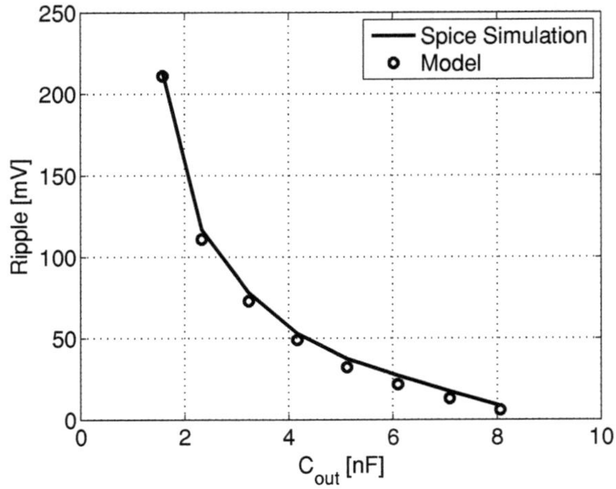

Fig. 4.5 Ripple simulation results of a fractional 4/5 converter: The converter has three flying capacitors, C_1 C_2 C_3, with respectively a capacitance of 0.25 0.25 and 0.5 nF. The switching frequency is 10 MHz

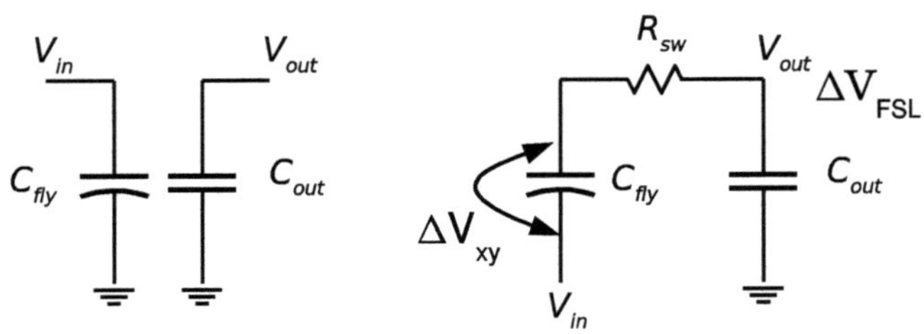

Fig. 4.6 Two states of a voltage doubler topology, including the switch resistance. The output noise is the filtered result of the flying capacitor's ripple

4.2.2 Noise in the Fast Switching Limit

For converters operating in the Fast Switching Limit, the parasitic resistance must be taken into account (Perigny et al. 2001). In order to demonstrate its influence a voltage doubler topology is analyzed.

If the voltage doubler in Fig. 4.6 is observed, the ripple can be approximated by a flying capacitor ripple $\Delta V_{x,y}$ damped by an RC-filter. The flying capacitor ripple (given a voltage doubler) can be approximated by:

$$\Delta V_{x,y} = \frac{I_{load}}{2 f_{sw} C_{fly}} \tag{4.4}$$

The transfer function of the RC filter, formed by the switch resistance R_{sw} and the buffer capacitor C_{out}, has an amplitude attenuation at the switching frequency of:

$$A_{RC} = \frac{1}{2 \times 2\pi f_{switch} RC_{OUT}} \tag{4.5}$$

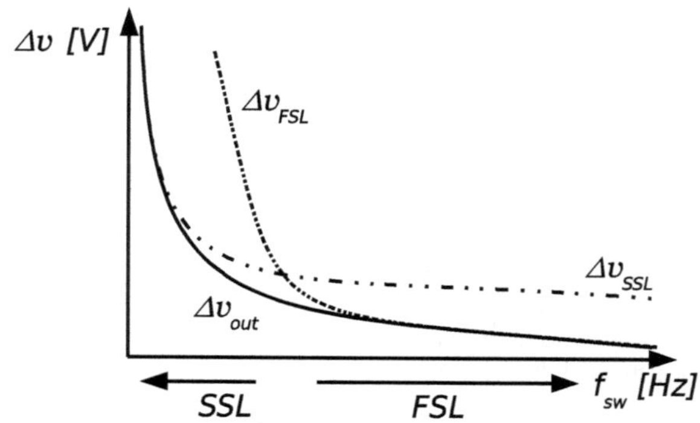

Fig. 4.7 Mathematical model of the voltage doubler's output voltage ripple

The resulting ripple at the output is approximated by:

$$\begin{aligned}\Delta v_{FSL} &= \Delta V_{x,y} A_{RC} \\ &= \frac{I_{load}}{8\pi f_{sw}{}^{2} R_{sw} C_{OUT} C_{fly}}\end{aligned} \tag{4.6}$$

This approximation suggests that the input signal is a sinusoidal signal at a frequency $2f_s$, while it would rather be a ramp-like signal and therefore overestimates the ripple. But this approximation gives a good idea of the ripple in FSL. Since most of the converters are operating in SSL, the SSL approximation is generally used.

In Fig. 4.7 both approximations of the ripple are shown for a voltage doubler. Obviously the SSL approximation only fits the observed ripple in the SSL dominated frequency range, while the FSL approximation is accurate at the higher end of the frequency range where the FSL effects act.

4.2.3 Additional Noise Sources

The noise contributions discussed in the previous paragraphs are clearly the predominant factors in the noise picture. However the following aspects should as well be taken into account: the parasitic resistance R_{ESR} of the output decoupling capacitors and the signal integrity of the control loop.

R_{ESR}-Noise

In Fig. 4.8 a generic model of a capacitor is shown. R_{leak} is the leakage resistance of the capacitor[2] and models the intrinsic discharging behavior of the capacitor.

[2] In this case only the charge leakage through the oxide is modeled. It is generally accepted that this leakage component is the dominant leakage phenomenon in CMOS integrated capacitors. (Rius and Meijer 2004).

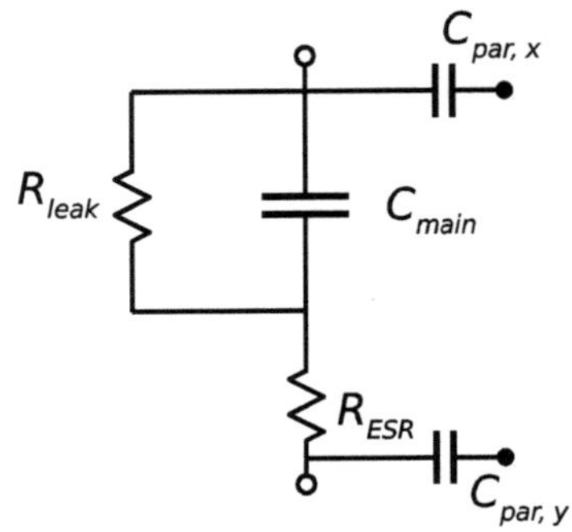

Fig. 4.8 Model of an integrated capacitor. The main functional capacitor is complemented with capacitors $C_{par,x/y}$ to include the effect of the capacitive coupling with the surrounding entities (substrate or power planes) and with an equivalent series/parallel resistance R_{ESR}/R_{leak} to fit the material properties of the capacitor

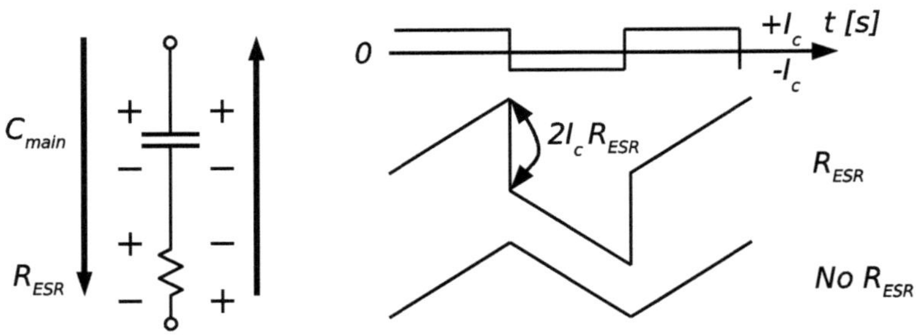

Fig. 4.9 Effect of the equivalent parasitic resistance of the capacitor on the capacitor voltage

The main capacitance C_{main} has also a series resistance R_{ESR}. It is this resistance which is problematic with respect to the output voltage noise.

A capacitor used as an output buffer (decoupling) capacitor is subject to alternating currents and even current pulses. An ideal capacitor, thus without R_{ESR}, can deal with this disregarding the current pattern. But when a series resistor is present, the voltage pattern across the capacitor and series resistor changes significantly.

In Fig. 4.9 is shown what happens if a constant but alternating current is applied to a realistic model of an output buffer capacitor. The voltage across the equivalent series resistor inverts at each change in current polarity. This invokes a voltage discontinuity over the capacitor in case that a parasitic series resistance is present.

But currents in capacitive converters are not constant, they are pulse-like, as is shown in Fig. 4.10. Again the voltage across the parasitic resistor changes polarity at once, which invokes a current spike which is superposed on the regular output voltage pattern. It is clear that the presence of a large equivalent parasitic series resistance invokes high-frequency content in the output voltage noise of the capacitive DC–DC converter. By proper design and selection of the integrated capacitors with low R_{ESR}, this issue can be omitted. The latter will be discussed in Chap. 6.

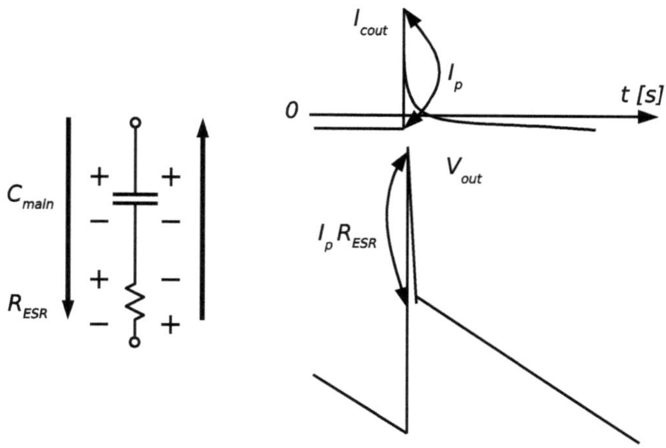

Fig. 4.10 Effect of the equivalent parasitic resistance of the capacitor on the capacitor voltage under influence of pulse-like current pattern

Signal Integrity of the Control Loop

In Chap. 5, a range of capacitive converter control techniques is presented. These control techniques are implemented by means of signal-processing blocks very much similar to the signal-processing block discussed in Sect. 4.1. It is thus of high importance to isolate this signal path from noise sources in the system. This can be achieved by means of proper layout and local decoupling.

4.3 Noise Power Loss

Scientific literature regards power-supply noise especially harmful from a signal quality point-of-view (Alon and Horowitz 2008; Stauth and Sanders 2007). But the following section also emphasizes the power loss due to a lack of noise mitigation. First from an analog circuit's point-of-view, next from a digital circuit's point-of-view. Keep in mind that most SoC's are mixed-signal by nature and thus both aspects have to be considered.

4.3.1 Analog Point-of-View

From an analog signal-processing point-of-view, supply noise is to be avoided since it corrupts the analog signals. Supply noise introduces errors in data converters (Senderowicz et al. 1997) or mixes with the power amplifier's frequency spectrum (Stauth and Sanders 2007). Especially the mixing of the DC–DC converter's switch-

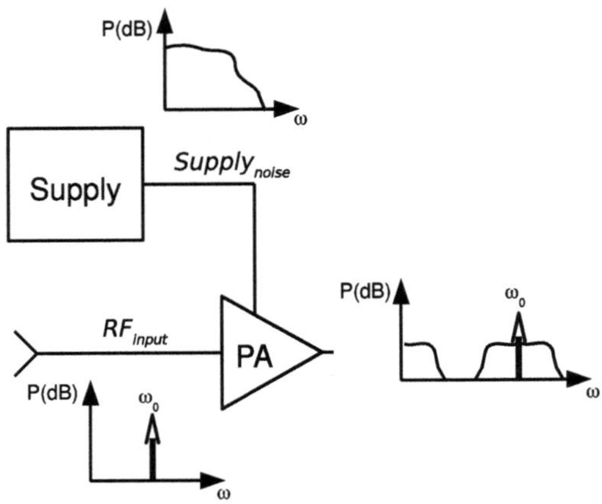

Fig. 4.11 Effect Modulation of an RF carrier with the supply noise in an RF Power Amplifier (PA)

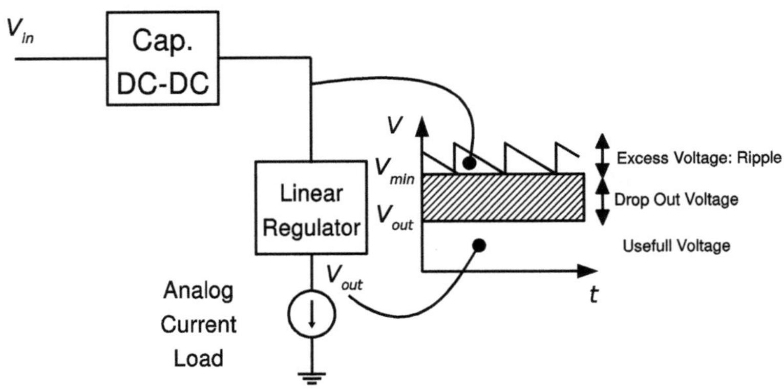

Fig. 4.12 Hybrid DC–DC conversion technique: the series regulator complements the capacitive DC–DC converter to improve the noise performance

ing frequency and the intended radio frequency is a detrimental effect in integrated transceivers. In Fig. 4.11 a power amplifier set-up is shown. This figure demonstrates how the power spectrum of the non-ideal power supply is mixed with the RF-carrier frequency ω_0, so that the base-band supply noise is shifted to the RF-carrier frequency. This effect potentially violates the required spectral mask specifications.

The most straightforward and widely used method to relax the DC–DC converters output voltage noise requirements, is the hybrid supply. This comes down to adding either a parallel (Stauth and Sanders 2007) or a series (MP1530) linear regulator to the switched-mode power supply. Due to its simplicity, the series-regulator approach is preferred.

In Fig. 4.12 the hybrid series-regulator approach is demonstrated. The set-up consists of a capacitive DC–DC converter with a series linear regulator. The capacitive

Fig. 4.13 From a digital circuit designer point-of-view, the system performance is determined by the minimum output voltage V_{min} of the supply. The ripple superposed on V_{min} introduces additional power loss

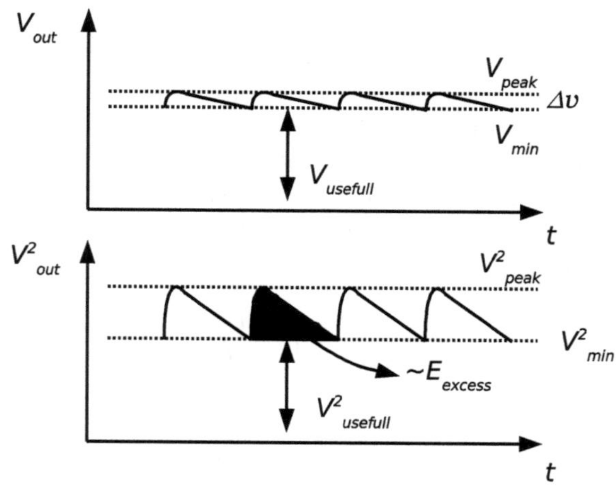

converter has an output buffer capacitance C^* and a switching period T. The linear regulator has a minimum input voltage V_{min} which corresponds to the minimum ripple voltage of the converter output and consumes part of the voltage range $V_{dropout}$. The linear regulator draws a constant current and thus loads the DC–DC converter with a constant current I_{load}. The power loss due to the linear regulator and the ripple is:

$$\Delta P = \frac{T I_{load}^2}{2C^*} + V_{dropout} I_{load} \tag{4.7}$$

$$= \frac{\Delta v_{SSL} I_{load}}{2} + V_{dropout} I_{load} \tag{4.8}$$

The first term involves the power loss due to the ripple and the second term originates from the linear regulator. The ripple loss can be minimized by decreasing the discharge time T or increasing the output capacitor's capacitance C^*. The second term can be minimized by reducing the drop-out voltage of the linear regulator.

4.3.2 Digital Point-of-View

Digital circuits are more robust to supply noise than analog circuits. These circuits only process voltages either close to the supply voltage or close to the ground voltage, leaving a large error margin on both sides, namely half of the supply voltage (Jan et al. 1996).

At first sight the noise originating from the DC–DC converter is only constrained by the noise margin of the digital circuits. But since the supply voltage influences the maximum speed of the circuit, a minimum supply voltage is determined. It is a good design practice to take the minimum supply voltage V_{min} as the lower boundary of the output voltage of the supply and the noise is superposed on this, as is shown in Fig. 4.13. The minimum input voltage also determines the useful power. Moreover the excess ripple—with a peak voltage V_{peak}—introduces a power loss with respect to the minimum required voltage. In Fig. 4.13 it is shown that the excess energy dissipated in the circuit is proportional to the marked area between $V_{out}^2(t)$ and V_{min}^2. The power loss due to the digital load with an equivalent resistance[3] R_{load} can be calculated as follows:

$$\Delta P = \frac{1}{\Delta v} \int_0^{\Delta v} \left(\frac{(V_{min} + v)^2}{R_{load}} - \frac{(V_{min})^2}{R_{load}} \right) dv \quad (4.9)$$

$$= \frac{1}{R_{load}} \left(V_{min} \Delta v + \frac{\Delta v^2}{3} \right) \quad (4.10)$$

4.4 Noise Mitigation Techniques

Both extrinsic and intrinsic techniques are used to reduce the noise level at the DC–DC converter's input/output. The series regulator approach is an extrinsic technique, since the noise reduction is achieved by adding a building block. But the multi-phase interleaving technique and the capacitance-modulation technique are intrinsic techniques. The noise reduction is accomplished by modifying the structure and the operation of the converter.

4.4.1 Series Regulator

A series regulator is an active dissipative system which ideally turns an unregulated (or poorly regulated) input voltage into a noise free DC output voltage.

In Fig. 4.14a a generic regulator is schematically represented. The excess voltage between input and output is dissipated in a resistance-like pass-device. To obtain an adequately regulated output voltage the resistance of the pass device is modulated to anticipate on input-voltage variations and to eliminate the input noise. Due to the dissipative nature of the device, a power loss and a voltage drop across the regulator's terminals occur.

The pass device is implemented by means of a MOS transistor operating in the triode region. Both an NMOS device or a PMOS device can be used. The choice between these devices is based on the trade-off between the device drop out and the ease of obtaining stability of the configuration. In Fig. 4.14b the configuration

[3] A digital load can be modeled as a capacitor switching between the supply and the ground and thus his behavior resembles a (switched-capacitor) resistor.

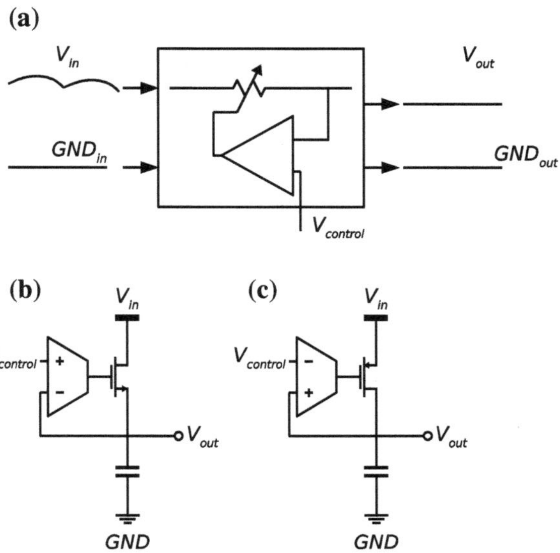

Fig. 4.14 **a** Generic interpretation of a series linear regulator. **b** NMOS-based linear regulator. **c** PMOS-based linear regulator or Low Drop Out Regulator

using an NMOS is shown. It has the disadvantage that the output voltage is at least one threshold voltage V_{th} lower than the maximum gate voltage of the pass device. In most implementations the maximum gate voltage is restricted to the input voltage, since the feedback circuitry is fed by the same input voltage as the pass device. The configuration in Fig. 4.14c uses a PMOS device and the voltage drop across the pass device is now function of the on-resistance of the PMOS device and the maximum load current. Thus the voltage drop across the PMOS pass device is typically much lower than the voltage drop across an NMOS pass-device.

Both configurations are closed-loop and hence stability of the loop is of high importance. At least two poles are present: one at the output node and one at the gate of the MOS pass device. The output pole is determined by the pass device's output impedance and the output capacitance, the internal pole by the amplifier's output impedance and the gate capacitance of the pass device.

The NMOS configuration has a low output impedance. In combination with the output capacitance this gives a high-frequency pole. The PMOS configuration has a high output impedance which depends on the load current and the low-frequency pole location shifts thus along with the load current.

The conventional compensation techniques can be applied but the varying pole location of the PMOS solution requires a worst case (dominant pole) compensation scheme. This requires a large compensation capacitance and is not favorable for integration purposes. Moreover until now only the two obvious poles in the system are observed, many more have to be taken into account in a realistic implementation. The amplifier has additional poles and the output capacitor has an equivalent series resistor (R_{esr}) which introduces a zero in the feedback loop. The latter makes the frequency behavior even more complex but can also be addressed for compensating one of the

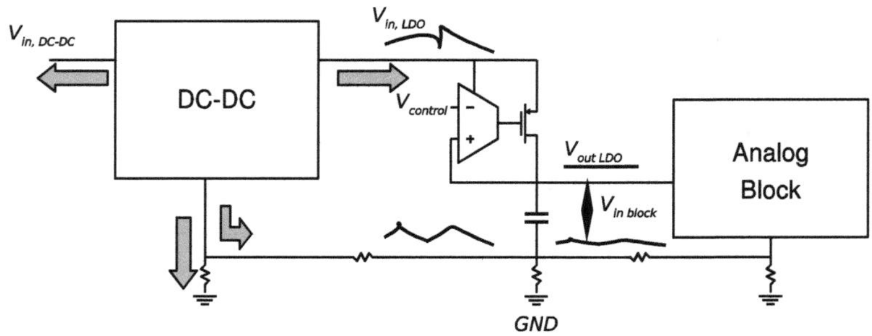

Fig. 4.15 Demonstration of converter-noise reduction by means of a linear regulator in a mixed signal SoC

poles. This requires complex compensation techniques and accurate characterization of the components.

Irrespective whether the PMOS or NMOS configuration is used, the DC–DC converter remains a heavy aggressor in the SoC. This is illustrated in Fig. 4.15: Although the linear regulator improves the quality of the supply voltage of the analog block, the DC–DC converter still injects charge or induces current pulses in the SoC substrate. These effects can be reduced[4] but not avoided. The only effective way is by reducing the noisy behavior at the source: modifying the DC–DC converter in such a way that the pulsed current patterns decrease in magnitude. The latter can be addressed by means of the following intrinsic techniques: multi-phase interleaving and capacitance modulation.

4.4.2 Multi-Phase Interleaving

For integrated converters it is very favorable to operate in the SSL. The SSL-ripple description gives also the upper limit for the ripple and thus a good starting point in the analysis of potential ripple reduction techniques. The ripple is calculated as follows:

$$\Delta v_{SSL} = \frac{I_{load}}{\zeta f_{sw}(C_{out} + \kappa_T C_{fly})} \tag{4.11}$$

The load current I_{load} is the only parameter in this equation which is determined by the specifications imposed by the application side. The remaining parameters are:

- The amount of output buffer capacitance C_{out}.
- The amount of flying capacitance C_{fly}.
- The discharge fraction ζ.

[4] By using separate ground connections and isolating the substrate by means of well's, pockets or guard rings.

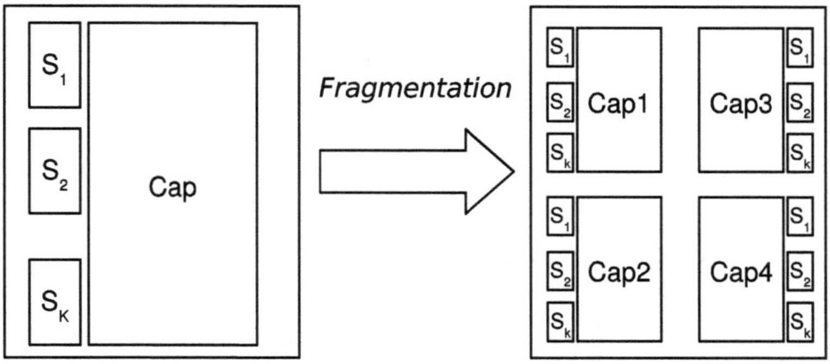

Fig. 4.16 Graphical representation of the fragmentation principle: Both capacitors and switches are reduced in size but increased in number. A single converter is replaced by N_{MP} smaller, but equivalent converter cores

- The switching frequency of the converter f_{sw}.

The most straightforward way to reduce the ripple is by increasing the amount of output decoupling capacitance C_{out}. As has been mentioned before the amount of capacitance is proportional to the chip area required for implementing the capacitor. So by increasing the output capacitor, the required chip area is increased without increasing the output power. Thus the power density of the DC–DC converter is reduced.

The switching frequency and the amount of flying capacitance can not be altered without penalizing the converter efficiency and κ_T depends on the converter topology.

This leaves us with the discharge fraction ζ to reduce the ripple. The following paragraphs demonstrate how to disconnect the discharge fraction ζ from the switching frequency.

Fragmentation

Fragmentation of a capacitive converter consists of replacing a single capacitive DC–DC converter by N_{MP} equivalent but N_{MP}-times smaller capacitive DC–DC converters. The switching frequency of each individual converter remains f_{sw}, the switches of each converter and the capacitor are scaled down with a factor N_{MP}. This is demonstrated in Fig. 4.16.

The switching pattern of a single capacitive DC–DC converter is shown in Fig. 4.17a. If the converter is fragmented in different converter cores as is shown in Fig. 4.16 and the cores are switched at the same frequency and in phase with each other: then the current pattern in Fig. 4.17b is observed. Although the current pattern of each individual capacitive converter is different[5] the total charge transfer is

[5] The peak current is reduced with a factor N_{MP} since the switch resistance increased by a factor N_{MP}.

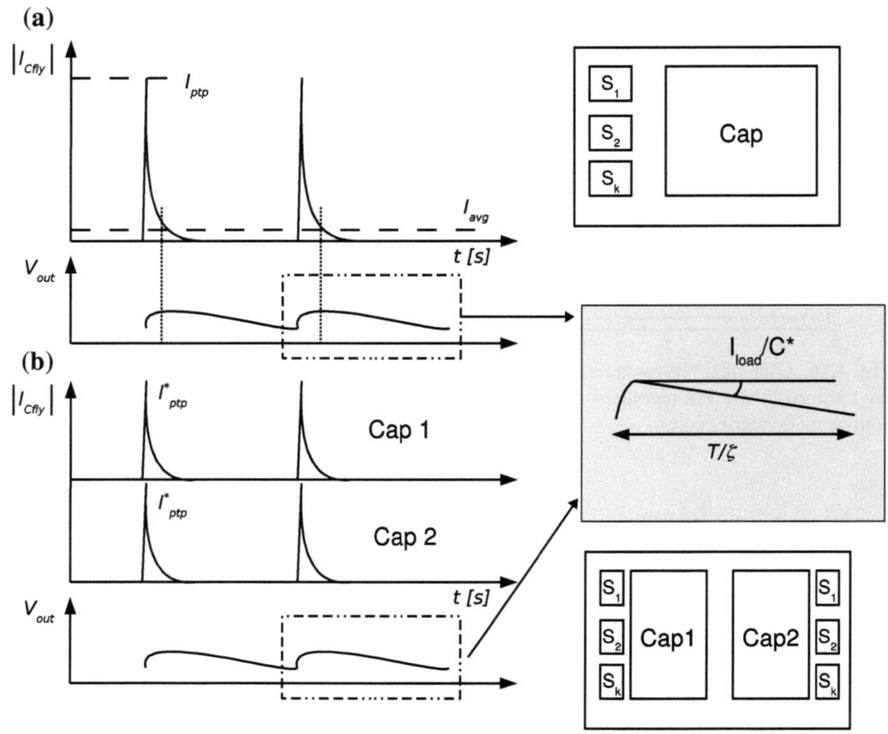

Fig. 4.17 **a** Current and voltage patterns of a single phase converter. **b** Current and voltage patterns of a fragmented converter, the different phases are switched in-phase

identical to the single converter approach and the output voltage pattern remains the same.

This approach assumes that the fragmentation of the converters has no impact on the components. It is demonstrated in Chap. 6 that both solid-state switches and capacitors can be fragmented without compromising their quality or characteristics.

Interleaving

In the previous paragraph it is claimed that fragmentation of a capacitive DC–DC converter does not influence the converter performance (power nor efficiency). Although fragmentation introduces an additional degree of freedom in the converter: the switching instant of each separate converter core. Instead of switching the different cores in phase, the cores can be shifted out of phase.

By doing so, the charge transfer is smeared out in time or formally said : the discharge fraction constant ζ is multiplied with a factor N_{MP} and the ripple is reduced with a factor N_{MP}:

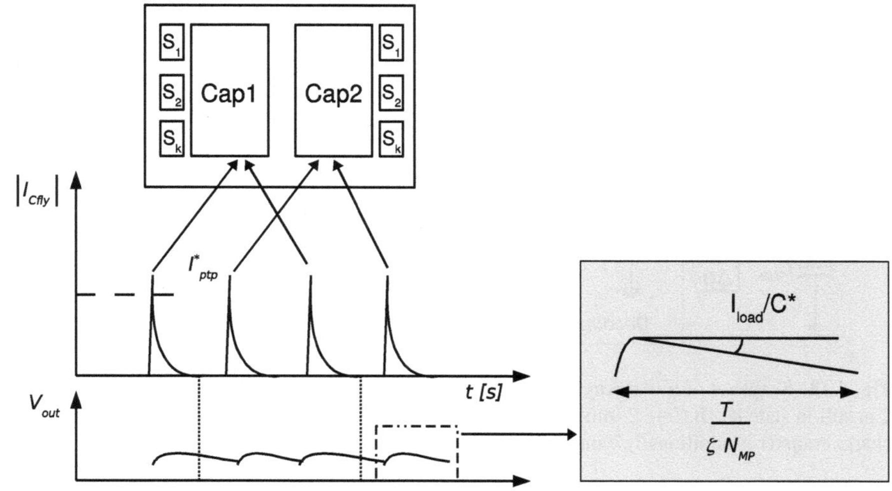

Fig. 4.18 Current and voltage patterns of a fragmented converter, the cores are switched in a multi-phase interleaved way

$$\Delta v_{SSL,MP} = \frac{I_{load}}{N_{MP}\zeta f_{sw}(C_{out} + \kappa_T C_{fly})} \tag{4.12}$$

In Fig. 4.18 it is demonstrated that only one core injects charge at a certain instant and during the time interval $\frac{T}{\zeta N_{MP}}$ associated with the active converter core, the other converter cores remain idle.

Additional Damping

The multi-phase approach introduces a supplementary benefit: additional damping by means of the idle cores. The idle cores are the converters cores which are not currently active, while they are connected between the output node and a DC node in the converter (in most cases the input node).

In Fig. 4.19 a snap shot of the two-phase interleaved voltage divider is shown. In the left pane the first core is transferring charge to the output while the second core still resides in the first state of the converter. In the right pane the second core initiates a charge transfer to the output node, while the first core is idle and remains in the second state. If these idle cores are connected between the output node and a DC node of the system, these cores will serve as additional damping when operating in the FSL region. Thus the idle cores are incorporated in C^* for the FSL ripple model.

Multi-phase interleaving has a second benefit besides the reduction of noise. Since the idle capacitors participate in the damping of the current pulses, they take over role of the output capacitor. In case a high level of interleaving is used, the dedicated output

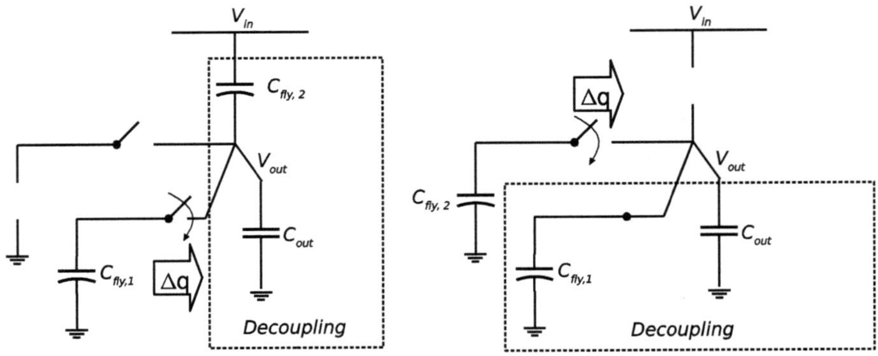

Fig. 4.19 Snapshot of a two-ways interleaved voltage divider: **a** core 1 is in state ϕ_1 while core 2 is still in state ϕ_2. **b** Core 2 initiates state ϕ_2 while core 1 remains in state ϕ_2. The active core's charge transfer is additionally damped by the idle cores

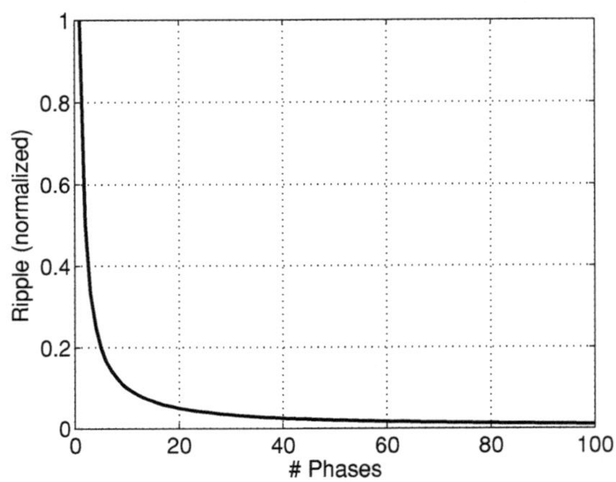

Fig. 4.20 Decreasing effect of multi-phase interleaving on the ripple amplitude. The ripple is normalized with respect to the ripple of a single-phase converter

capacitor can be omitted. This provides a significant increase in power density since the output capacitor typically takes a lot of area without adding any output power.

Figure 4.20 shows the normalized ripple as a function of the number of interleaving phases. It is clear that around twenty phases the impact of an additional phase can be neglected. Moreover in Chap. 6 it is demonstrated that ongoing fragmentation has a negative impact on the quality of the capacitors. Both arguments are taken into consideration when determining the level of interleaving.

4.4.3 Capacitance Modulation by Means of Fragmentation

The multi-phase interleaving technique addresses the parameter ζ as a tool to reduce the switching noise or ripple at the converter's output node. Another technique is

Fig. 4.21 Heavy load current and voltage patterns of: **a** A single capacitive converter. **b** A fragmented capacitive converter switched in-phase

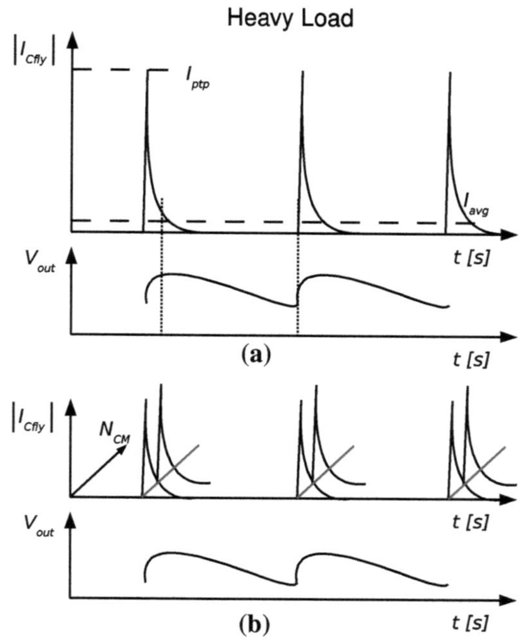

to increase the switching frequency for a fixed load current (I_{load}). But the output power and the output current are proportional to the switching frequency and the amount of flying capacitance. Thus a change in frequency must be compensated by an opposite change in flying capacitance. This can be achieved by fragmenting the converter in N_{CM} equivalent cores. By doing so a fraction of the converter can be selected and a change in frequency can be compensated by a change in core activity.

This approach is infeasible at the maximum load point of the converter, since an increase in switching frequency drives the converter into FSL, which eliminates the frequency sensitivity of the converter. The capacitance-modulation technique is thus especially beneficial at light loads. But it is especially at these low loads that the ripple of capacitive converters increases due to the reduced impact of the FSL. Therefore capacitance modulation is an effective technique to reduce the low-load ripple excursions.

In Fig. 4.21 the heavy-load condition is demonstrated. Panel (a) represents the single converter approach panel (b) the fragmented converter. It is clear that there is no effect on the output voltage if the fragmented cores switch at the same instant. The equivalent charge transfer and the total current pulse are identical.

In Fig. 4.22 the load is halved, instead of reducing the switching frequency by the same factor as is done by the single converter solution (panel a), the fragmented converter deactivates half of the converter cores, but the remaining cores maintain the same switching frequency as under heavy-load conditions.

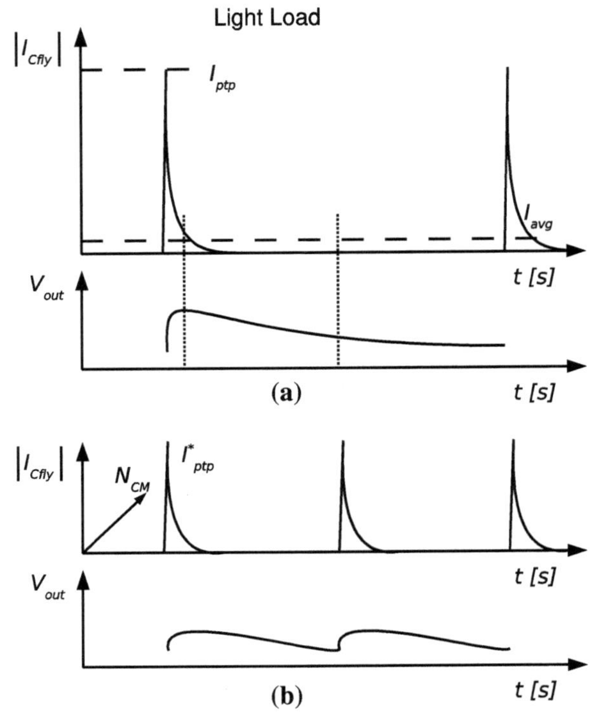

Fig. 4.22 Light load current and voltage patterns of: **a** Single capacitive converter. **b** Fragmented capacitive converter switched in-phase

4.5 Conclusion

This chapter identifies a fully integrated capacitive DC–DC converter as a potential noisy aggressor in a fully grown SoC. It is of primary importance to control the parasitic paths and to model the unwanted coupling between the different building blocks in a SoC. But even more important is it to deal with the issue by avoiding the noise injection at the source : the converter itself. Therefore three techniques have been analyzed. The linear regulator is a good way to reduce the noise on the direct power supply connection but can not eliminate the noise that will migrate by means of the parasitic connections. The multi-phase interleaving technique and the capacitance-modulation technique deal with the root of the problem: the high-amplitude current pulses. In fact by modifying the converter operation the current magnitude is decreased which results in both cases in a ripple smeared out in time. These techniques provide an adequate set of tools to reduce the noise levels originating from a fully integrated DC–DC converter.

Chapter 5
Control of Fully Integrated Capacitive Converters

Control of DC–DC converters refers to the manipulation of the converter's parameters to match the DC–DC converter's behavior with the system supply requirements. These system supply requirements are imposed either through the power-source characteristics from Sect. 1.1 (Chap. 1) or through power-management specifications, discussed in Sect. 1.2. Irrespective of the motivation for using a DC–DC converter, the primary concern is to ensure a predictable output voltage and a predictable response on both predictable and unforeseen perturbations.

This chapter looks at the capacitive DC–DC converter from a totally different point of view: how to design a DC–DC converter that facilitates the best possible control? The *best possible control* is the behavior that meets the requirements imposed by the system level and achieve this at the lowest cost.[1] To meet these requirements this chapter provides an overview of system-analysis methods and control techniques. In Sect. 5.1 the nature of a capacitive DC–DC converter is reviewed with respect to its controllability. Next the analysis methods for capacitive converters are elaborated in Sect. 5.2. In Sect. 5.3 the most important control techniques: Topology Reconfiguration (TR), Pulse-Width Modulation (PWM) and Pulse-Frequency Modulation (PFM) are analyzed and finally Sect. 5.4 covers their high-level implementation of these techniques. Based on this assessment conclusions are drawn in the final section.

5.1 Control Nature

The perfect DC–DC converter behaves like an ideal control-dependent DC–DC transformer. It has neither output nor input impedance, guarantees power conservation and its transformation ratio is controllable. This results in a circuit that supplies any current and any voltage with zero delay and error with respect to the control reference. The control reference is an externally fed signal, it can be digital or analog, which

[1] In IC design cost is quantified in IC area and/or power consumption.

T. Van Breussegem and M. Steyaert, *CMOS Integrated Capacitive DC–DC Converters*, Analog Circuits and Signal Processing, DOI: 10.1007/978-1-4614-4280-6_5,

determines the output voltage. In most cases there is a known relationship between the output voltage and the control reference voltage (often a linear relationship).

Control in a Power-Management Architecture:

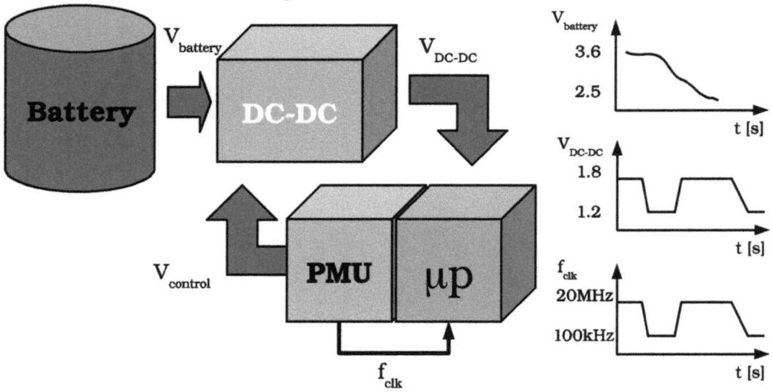

Fully integrated DC–DC converters can be used to provide Dynamic Voltage Scaling (DVS) in a SoC, as is discussed in Chap. 2. The latter is demonstrated in the above setup, for a digital functional block. DVS includes the ability to deliver a constant high voltage when the digital block is operating at high clock frequencies in a so called high-performance mode, but equally to change the voltage to a much lower level when the digital block enters a low-power mode and experiences a lower clock frequency. The DC–DC converter must be able to execute this transition within the required time period (state of the art: $<10\,\mu s$) and at the same time deal with a large change (in some applications more than two decades) in delivered output power. Moreover, in a good power-management interface the output voltage is kept under control irrespective of the input voltage. In case that a battery is used as an energy source, the input voltage is subject to large changes in amplitude due to the relationship between the output resistance of the battery and the amount of energy stored in the battery.

In Chap. 2 the output impedance model for capacitive DC–DC converters has been introduced. The output impedance R_{out} included in this model is non-zero and depends on the converter characteristics and the frequency or duty cycle of the switching frequency. The output impedance model imposes a first set of obstacles to design a capacitive converter while meeting the perfect power supply requirements as is demonstrated in Fig. 5.1. Due to the non-zero output impedance, the capacitive converter can not operate without power loss. Moreover, the output voltage V_{out} is a function of the load current if the switching frequency and the duty cycle of the switching frequency are constant. The transformation ratio (N) is fixed by means of the topology of the configuration.

These issues are addressed by the control block. The control block is basically a signal-processing block, continuous or discontinuous by nature, which observes the

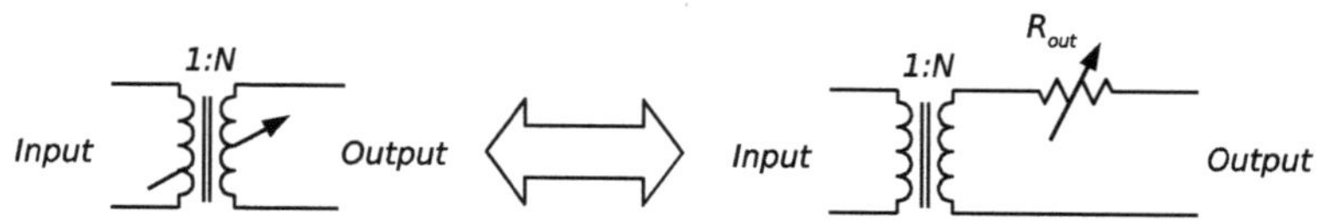

Fig. 5.1 Comparison between an ideal power supply and the equivalent output impedance model of a capacitive DC–DC converter

DC–DC converter's behavior, tunes it to meet the system requirements and mimics an ideal DC–DC transformer.

There exists a strong association between the range of potential control methods and the nature of the capacitive DC–DC converter. In fact the control methods are readily derived from the output impedance model. The output voltage is altered either by changing the ideal VCR (N) or by changing the value of the output impedance R_{out}. The only way to change the VCR is by implementing an alternative topology, this technique is called Topology Reconfiguration (TR). For controlling the output impedance primarily Pulse-Frequency Modulation (PFM) and Pulse-Width Modulation (PWM) are used. The following sections cover a number of control methods, but first the capacitive DC–DC converter's frequency-domain behavior is analyzed.

5.2 Frequency-Domain Analysis

A DC–DC converter operating in the FSL can be controlled by changing the switching frequency's duty cycle, in literature this technique is known as Pulse Width Modulation (PWM). A converter operating with a strong SSL influence, controls the output impedance by changing the switching frequency by a so-called Pulse Frequency Modulation (PFM) scheme or by modifying the amount of flying capacitance C_{fly}. Both SSL and FSL require a different analysis technique of the converter, this is discussed in the following paragraphs.[2]

5.2.1 Frequency-Domain Analysis in FSL

Conventional frequency-domain analysis of DC–DC converters is based on the Time-Averaging Approach (TAA) (Cuk 1976; Krein et al. 1990). Since the discussed DC–DC converters are dual-state VSS's, each one of the converter states is described by means of a different set of equations and thus delivers a composite set of equations.

[2] Alternative control techniques, for example switch-resistance modulation, are omitted in this chapter. We focus on the dominant control techniques especially since the derived mathematical methods can be addressed for analyzing the other techniques as well.

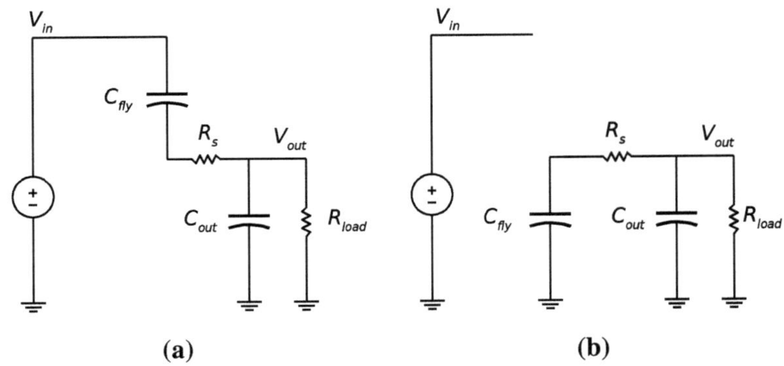

Fig. 5.2 Both configurations of a two-state capacitive DC–DC converter: the series parallel divide-by-two. **a** State ϕ 1 with the flying capacitor between input and output node; **b** state ϕ 2 with the flying capacitor between ground and output node

The TAA suggests that a single set of equations can be extracted by calculating a time-weighted average from the composite set. The resulting approximation is only valid if the system's time constants are much larger than the switching period of the VSS (Cuk 1976).

As discussed in Chap. 2 the excessive time constants are a prerequisite for capacitive converters to operate in the FSL. Thus TAA can be addressed to describe the frequency-domain behavior of capacitive DC–DC converters operating in the FSL.

In order to demonstrate this technique, the frequency-domain response of a two-state divide-by-two series-parallel capacitive converter, depicted in Fig. 5.2, is calculated. The resistance R_s represents the equivalent aggregate resistance of the switches[3] and the parasitic resistance of the capacitor. This VSS has a two-state nature and each state is described by a set of differential equations:

State ϕ 1:

$$C_{fly}\frac{dV_{C_{fly}}}{dt} - C_{out}\frac{dV_{C_{out}}}{dt} = \frac{V_{C_{out}}}{R_L} \tag{5.1}$$

$$V_{C_{fly}} = V_{in} - V_{C_{out}} - R_s C_{fly}\frac{dV_{C_{fly}}}{dt} \tag{5.2}$$

[3] It is assumed that the switch resistance is independent from the output and input/output voltage and constant during the entire state. Violating these prerequisites results in a non-linear set of equations.

State ϕ 2:

$$-C_{fly}\frac{dV_{C_{fly}}}{dt} - C_{out}\frac{dV_{C_{out}}}{dt} = \frac{V_{C_{out}}}{R_L} \tag{5.3}$$

$$V_{C_{fly}} = V_{C_{out}} - R_s C_{fly}\frac{dV_{C_{fly}}}{dt} \tag{5.4}$$

Continuous dynamic systems are modeled by means of a state-space model. The state variables $V_{C_{fly}}$ and $V_{C_{out}}$ fully determine the operation of the converter in a unique way and enable thorough analysis of the system dynamics. The system's differential equations are decoupled in order to comply with the general form of the general state matrix in Eqs. 5.5–5.6.

$$\dot{\vec{X}} = A\vec{X} + B\vec{U} \tag{5.5}$$

$$\vec{X} = \left(V_{C_{fly}} V_{out}\right)^T \tag{5.6}$$

State ϕ 1:

$$\frac{dV_{C_{fly}}}{dt} = \frac{V_{in} - V_{C_{out}} - V_{C_{fly}}}{R_s C_{fly}} \tag{5.7}$$

$$\frac{dV_{C_{out}}}{dt} = -\frac{V_{C_{fly}}}{R_s C_{out}} - \left(\frac{1}{R_s C_{out}} + \frac{1}{R_L C_{out}}\right)V_{C_{out}} + \frac{V_{in}}{R_s C_{out}} \tag{5.8}$$

State ϕ 2:

$$\frac{dV_{C_{fly}}}{dt} = \frac{V_{C_{out}} - V_{C_{fly}}}{R_s C_{fly}} \tag{5.9}$$

$$\frac{dV_{C_{out}}}{dt} = \frac{V_{C_{fly}}}{R_s C_{out}} - \left(\frac{1}{R_s C_{out}} + \frac{1}{R_L C_{out}}\right)V_{C_{out}} \tag{5.10}$$

Next a single set of equations, describing the time derivative of the capacitor voltages, is calculated by averaging the state equations over a switch period T.

$$\frac{dV_{C_{fly}}}{dt} = \frac{-1}{R_s C_{fly}}V_{C_{fly}} + \frac{1-2D}{R_s C_{fly}}V_{C_{out}} + \frac{D}{R_s C_{fly}}V_{in} \tag{5.11}$$

$$\frac{dV_{C_{out}}}{dt} = \frac{1-2D}{R_s C_{out}}V_{C_{fly}} - \left(\frac{1}{R_s C_{out}} + \frac{1}{R_L C_{out}}\right)V_{C_{out}} + \frac{D}{R_s C_{out}}V_{in} \tag{5.12}$$

The following characteristic radial frequencies are observed: $\omega_1 = \frac{1}{R_s C_{fly}}$ $\omega_2 = \frac{1}{R_s C_{out}}$ $\omega_3 = \frac{1}{R_L C_{out}}$. This reduces the equations to:

$$\frac{dV_{C_{fly}}}{dt} = \omega_1 V_{C_{fly}} + (1-2D)\omega_1 V_{C_{out}} + \omega_1 D V_{in} \tag{5.13}$$

$$\frac{dV_{C_{out}}}{dt} = (1-2D)\omega_2 V_{C_{fly}} - (\omega_2+\omega_3) V_{C_{out}} + D\omega_2 V_{in} \tag{5.14}$$

These equations are linearized around an operation point ($\overline{V_{Cfly}}\ \overline{V_{Cout}}\ D$) the following equation is obtained:

$$d\frac{\overline{V_{C_{fly}}} + v\tilde{C}_{fly}}{dt} = \omega_1(\overline{V_{C_{fly}}} + v\tilde{C}_{fly}) + (1-2D-2\tilde{d})\omega_1(\overline{V_{C_{out}}} + v\tilde{C}_{out}) + \omega_1(D+\tilde{d})(\overline{V_{in}} + \tilde{v_{in}}) \tag{5.15}$$

$$d\frac{V_{C_{out}} + v\tilde{C}_{out}}{dt} = (1-2D-2\tilde{d})\omega_2(\overline{V_{C_{fly}}} + v\tilde{C}_{fly}) - (\omega_2+\omega_3)(\overline{V_{C_{out}}} + v\tilde{C}_{out}) + (D+\tilde{d})\omega_2(\overline{V_{in}} + \tilde{v_{in}}) \tag{5.16}$$

Taken that the derivatives of a constant (operation point) are zero and there is no variation in input voltage, the ensuing equations describe the dependency on small signal perturbations:

$$d\frac{v\tilde{C}_{fly}}{dt} = \omega_1(\overline{V_{C_{fly}}} + v\tilde{C}_{fly}) + (1-2D-2\tilde{d})\omega_1(\overline{V_{C_{out}}} + v\tilde{C}_{out}) + \omega_1(D+\tilde{d})(\overline{V_{in}}) \tag{5.17}$$

$$d\frac{v\tilde{C}_{out}}{dt} = (1-2D-2\tilde{d})\omega_2(\overline{V_{C_{fly}}} + v\tilde{C}_{fly}) - (\omega_2+\omega_3)(\overline{V_{C_{out}}} + v\tilde{C}_{out}) + (D+\tilde{d})\omega_2(\overline{V_{in}}) \tag{5.18}$$

A Laplace transformation and substitution of the flying capacitor voltage results in the frequency-domain transfer-function:

$$\frac{v\tilde{C}_{out}}{\tilde{d}} = \frac{-2\omega_2\overline{V_{C_{fly}}} + (1-2D)\omega_2\frac{-2\omega_1\overline{V_{C_{out}}}+\omega_1\overline{V_{in}}}{s-\omega_1} + \omega_2\overline{V_{in}}}{s+(\omega_2+\omega_4) - \frac{(1-2D)\omega_2}{s-\omega_1}} \tag{5.19}$$

The transfer function from variations in duty cycle $\tilde{d}$ to variations in output voltage $v\tilde{C}_{out}$ describes a two-pole system with a single zero. Similarly to inductive type converters this results in complex frequency-compensation schemes to anticipate on both the operation point and the load-condition dependency of the pole and the zero-locations. Moreover the compensation scheme only fits a converter operating in FSL. Therefore a fixed switching frequency is required. This results in a fixed switching power loss irrespective of the converter's output power, as is shown in Fig. 5.3a, and thus a degradation in conversion efficiency for relatively low output-power conditions. For that reason an SSL-mode converter is preferred, so that a PFM

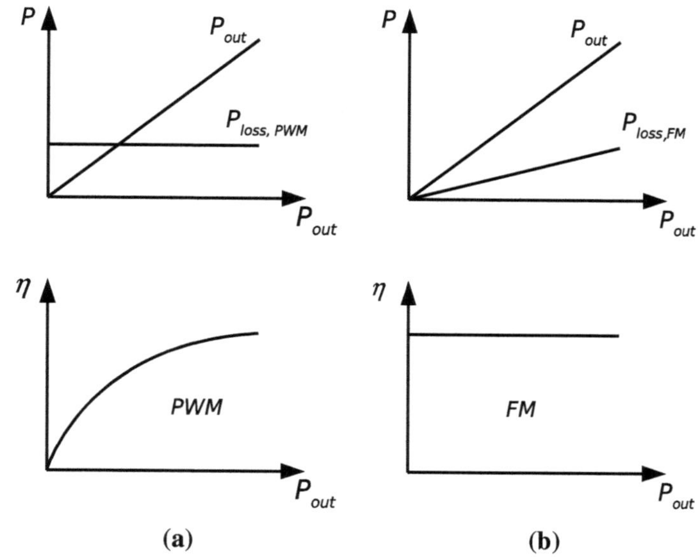

Fig. 5.3 Representation of the power loss $P_{loss,xxx}$ due to PWM (**a**) and PFM (**b**), with respect to the total output power P_{OUT} and the effect of this on the overall system efficiency η

control can be implemented. This results in a switching-power loss that scales with the output power as is demonstrated in Fig. 5.3b.

But the frequency response also reveals that a linearization around duty cycle $D = 0.5$ results in a single pole description and eliminates the zero all together:

$$\frac{\tilde{v_{C_{out}}}}{\tilde{d}} = \frac{f_2\overline{V_{in}} - 2f_2\overline{V_{C_{fly}}}}{s + f_4} \tag{5.20}$$

This suggest that fixed duty-cycle control, as is imposed by PFM, results in a less complex frequency-domain behavior.

5.2.2 Frequency-Domain Analysis in SSL

The TAA can not be used since by definition the capacitive converter's time constants are small with respect to the converter's switching period when operating in SSL. But in Chap. 2, the charge-balance method is described as an alternative method to synthesize the input to output transfer function of a capacitive DC–DC converter. In order to design a switching frequency-controlled capacitive DC–DC converter, the frequency-to-output voltage transfer function is required. Based on the methodology in Chap. 2, the discrete-time charge-balance method with varying switching period is obtained:

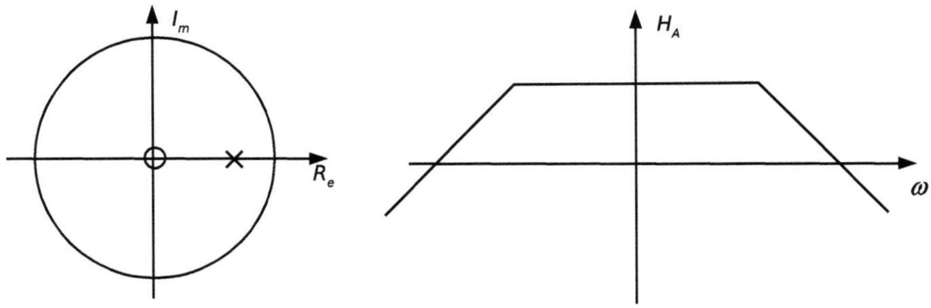

Fig. 5.4 The Bode plot of a divide-by-2 capacitive DC–DC converter operating in SSL and controlled by a PFM-like control technique

$$C_{fly}(V_{in} - V_{out}(n-1)) + \frac{I_{load}T(n-1)}{2} - \frac{I_{load}T(n-1)}{2} + C_{out}V_{out}(n-1)$$
$$= (C_{fly} + C_{out})V_{out}(n) + \frac{I_{load}T(n)}{2} \tag{5.21}$$

Straightforward transformation ($t \rightarrow Z$) to the frequency domain, while neglecting the DC terms, gives us the following system transfer function:

$$\frac{V_{out}}{T} = \frac{-I_{load}}{2} \frac{1}{(C_{fly} + C_{out}) - (C_{out} - C_{fly})z^{-1}} \tag{5.22}$$

This is a single-pole system, shown in Fig. 5.4, with a low-pass-like Bode plot. Similarly to the analysis of the FSL frequency response, the remaining pole location is a function of the converter's operating point. However single-pole systems have a much larger phase margin and are stable by nature. The latter promotes converters operating in SSL and controlled by a Pulse-Frequency Modulation technique.

This modeling approach is verified by extracting the frequency response of a voltage divider by simulation in Spice and validating the discrete-time model. In Fig. 5.5 the continuous line represents the Matlab approximation of the discrete-time frequency response, the simulation data is added by means of the black squares. In fact a good approximation of the simulation data is obtained. In Table 5.1 the characteristics of the capacitive DC–DC converter, which is used in this verification, are found. This approximation is only valid in case that the output capacitor is relatively large with respect to the flying capacitor and in case that the pole is positioned far enough from the switching frequency.

5.3 Techniques

In order to control a DC–DC converter, a signal-processing block is used. This block observes the system characteristics and intervenes in the converter's behavior to comply with the a-priori defined conversion characteristics. These charac-

Table 5.1 Converter characteristics for frequency domain validation

$iVCR$	1/2
V_{in}	2 V
V_{out}	0.7 V
C_{fly}	500 pF
C_{out}	10 nF
R_{load}	1000 Ω

teristics can include the converter's output voltage, the output current, the settling time, the start-up time, start-up overshoot voltage and many more. The most important DC–DC converter characteristic is the output voltage and the variation of the output voltage under influence of a change in load condition (the so called load-regulation).

The previous section issued the dominant methods for analyzing capacitive DC–DC converters. The use of methods based on duty cycle modulation is strongly discouraged in low power fully-integrated converters. Especially due to the detrimental impact of the constant switching losses on the conversion efficiency in case of low load conditions. The previous section also demonstrates the multi-pole nature of a capacitive converter using duty cycle modulation (PWM). Frequency-modulation techniques are appreciated especially for their straightforward implementation and less complex frequency-domain nature. If a constant switching frequency is required, Capacitance Modulation is an appropriate alternative for the PWM technique. In this section a selection of control techniques is presented.

5.3.1 Topology Reconfiguration

In previous chapters the tight relationship between the voltage conversion range and the topology of a capacitive DC–DC converter, is highlighted. Every topology is associated with a fixed iVCR and only for this VCR the converter operates at 100 % efficiency. The iVCR can not be exceeded and if the ratio is smaller this comes at the cost of efficiency η since:

$$\eta_{max} = \frac{V_{out}}{N V_{in}} \tag{5.23}$$

$$= \frac{V_{out}}{V_{out,ideal}} = \gamma \tag{5.24}$$

Large excursions in VCR are penalized by means of a degradation of the conversion efficiency. Therefore it is of high importance to match the topologies VCR with the required output voltage. In case that a large input-output range is required, it is beneficial to opt for a capacitive DC–DC converter comprising multiple topologies.

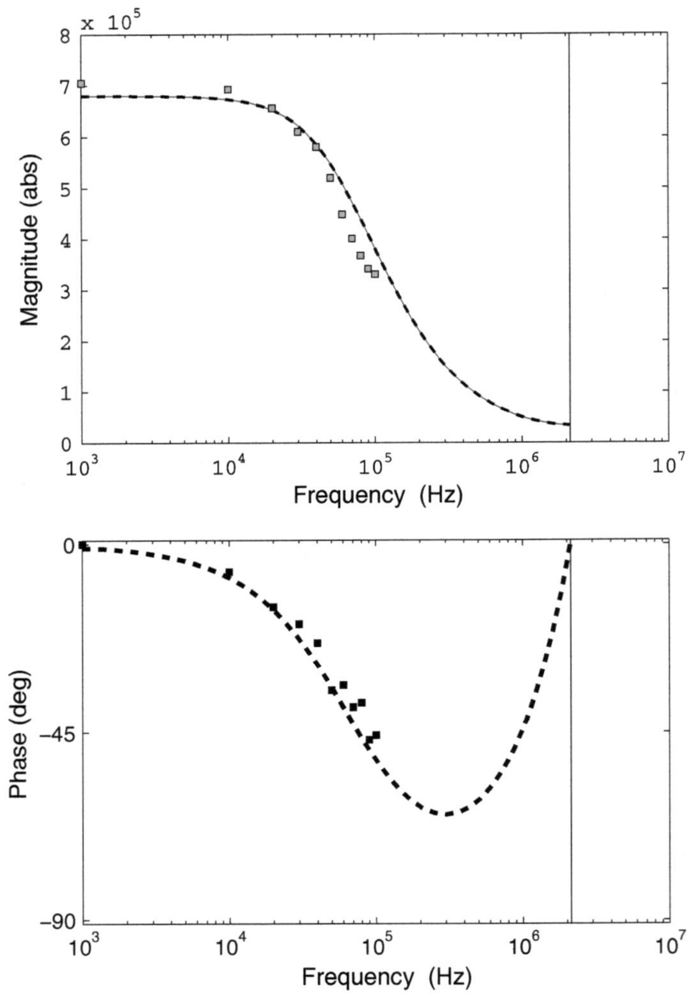

Fig. 5.5 Validation of the discrete-time transfer function of the voltage divider with respect to spice simulations

This implies that the converter is reconfigured to match the input-output ratio with the optimum iVCR of the topology set as has been discussed in Chap. 2.

In Fig. 5.6 a combined 2/3 and 4/5 fractional capacitive converter is depicted. Adding a topology boosts the efficiency for a certain fraction of the input-output range, but also influences the performance on the remainder of the range. The influence of multiple topologies in a single converter is analyzed in Chap. 3 and is identified to have a non-negligible impact on the overall performance. This effect is demonstrated in Fig. 5.7. Boosting the efficiency at the lower VCR-end by adding an additional topology , depresses the efficiency at the originally higher VCR-end.

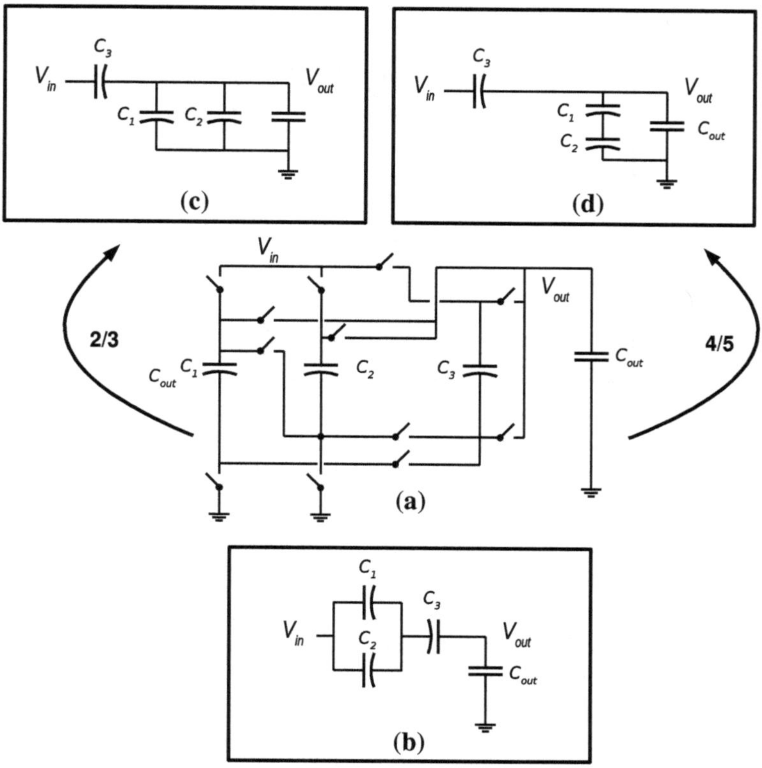

Fig. 5.6 This structure (**a**) denotes a multiple-topology converter. A fixed set of capacitors and switches is used for addressing a 4/5 topology (**d**) and a 2/3 topology (**c**). These topologies are both characterized by a common phase, demonstrated in (**b**)

This effect can be compensated by compromising the power density. Since topology reconfiguration only manipulates the transformer ratio in the Output Impedance Model, additional techniques remain necessary to deal with load-current variations. Hence TR is mainly used to address line regulation.

5.3.2 Capacitance Modulation

The output impedance of a capacitive DC–DC converter operating in the SSL can also be changed by modifying the amount of flying capacitance C_{fly}. A capacitive converter is equally sensitive to a variation of flying capacitance as it is to a variation in switching frequency.

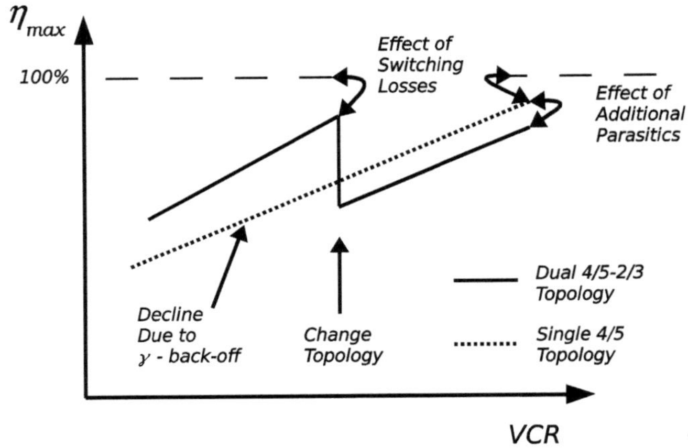

Fig. 5.7 Boosting the efficiency at the lower VCR-end, by adding an additional topology, reduces the efficiency at the originally higher VCR-end

Until recently topology reconfiguration was rarely used in conventional DC–DC converters, but due to the monolithic integration opportunities this is even considered for inductive converters (Zheng and Ma 2010). The main obstacle for topology reconfiguration is the immanent strive for reducing the component count (based on cost) and the system's footprint (based on both cost and convenience). Thanks to the advent of integration, discussed in Chap. 1, the number of components is of no issue any more, although the circuit size remains critical.

Therefore, building in reconfigurability is mainly considered if this requires minimum impact of the converter's power density (output power with respect to converter area). Considering both the state of the art as the predictions of CMOS technology, the integrated capacitors remain the dominant area consumers. Therefore in case of fully integrated capacitive DC–DC converters, reconfigurability is built in without a significant power density penalty.

Applying this technique requires a method to modify the amount of flying capacitance involved in the conversion. The latter can be achieved by fragmenting the converter in multiple equivalent converter cores. Similarly to the approach followed in the multi-phase technique, the switches are scaled accordingly. By deactivating a fraction of the cores the total equivalent output impedance of the converter is altered while keeping the switching frequency constant. In Fig. 5.8 the equivalent model of a fragmented capacitance-modulated capacitive DC–DC converter is demonstrated. The figure shows N_{CM} paralleled converters that can be activated individually according to the required output power. Instead of manipulating the switching frequency, the number of active converter-cores is adjusted.

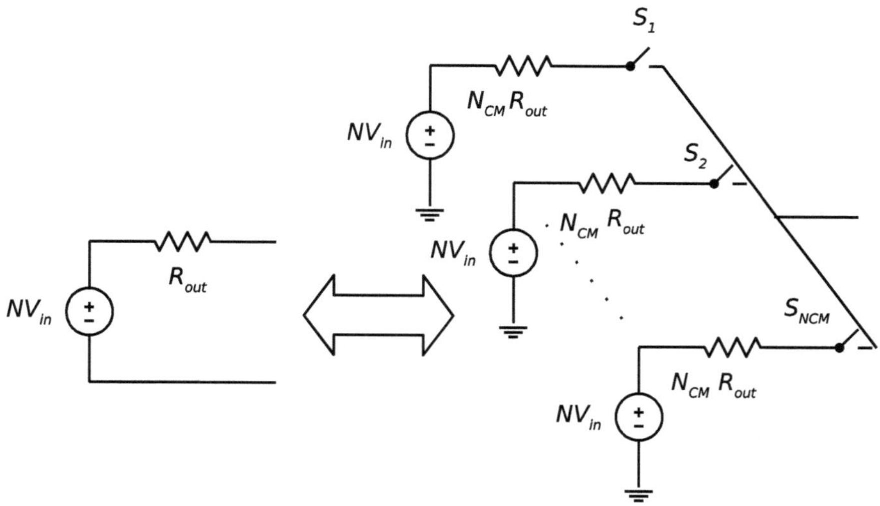

Fig. 5.8 Equivalent model of a fragmented capacitance modulated capacitive DC–DC converter

5.3.3 Pulse-Width Modulation

PWM is a wide-spread technique used in power electronics as well as in many other fields. In capacitive DC–DC converter design it is used as a straightforward technique to alter the converter's output impedance. But it is equally applicable to control capacitive DC–DC converters operating in the FSL.[4]

In Chap. 3 it is demonstrated by Eq. 3.10, that the output voltage V_{out} can be calculated as the resistive division between the output impedance R_{out} and the equivalent load resistance R_{load}:

$$V_{out} = NV_{in}\frac{R_{load}}{R_{load} + R_{out}} \tag{5.25}$$

This suggests that the most straightforward way to anticipate on load (resistance) variations is modifying the output impedance of the converter. For a fixed input voltage the output voltage is kept constant by fixing the equivalent voltage drop across the output impedance. If the load current changes, the output impedance must track this and the control circuitry keeps the output voltage constant. For a capacitive converter operating in the FSL the output impedance is determined by the switch conductance G_i, the topologies charge switch vectors $a_{s,i}$ and the duty cycle D of the switching frequency. Therefore the duty cycle D can be used to modify the converter's output impedance, which is demonstrated in Fig. 5.9a. By doing this the effect of the output impedance can be masked and the output voltage kept constant.

[4] In the SSL the output impedance is insensitive to duty-cycle variation, this has been discussed in Chap. 2.

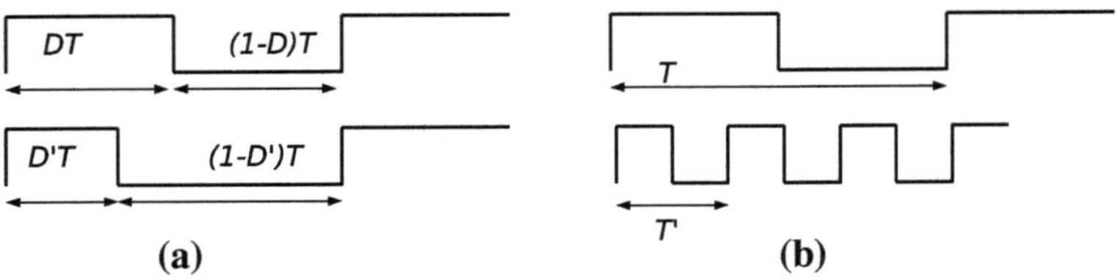

Fig. 5.9 **a** Pulse-Width Modulation, changes the switching frequency's duty cycle; **b** Pulse-Frequency Modulation operates on the period of the switching frequency

The fixed switching frequency and thus fixed switching losses have a negative impact on the converter's efficiency at the low output-power end. Therefore PWM is hardly used in the field of fully integrated DC–DC converters and therefore not further discussed.[5]

5.3.4 Pulse-Frequency Modulation

A converter operating in the SSL has an output impedance which is mainly determined by the inverse of the switching frequency f_{sw} and the amount of total flying capacitance C_{fly}. Thus control of the output voltage can be achieved by modifying the output impedance value according to the load current. Modulating the switching frequency, this is demonstrated in Fig. 5.9b, is the most straightforward approach to change the output impedance in SSL. The main advantage with respect to PWM is that for PFM the switching frequency scales proportionally along with the output power. This means that the efficiency does not roll off for low loads as it does for PWM.

Continuous Time Control

In Eq. 5.22, the main transfer function of a capacitive converter, operating in the SSL, is proposed. Both demonstrating the single dominant pole nature of the converter. In order to control the output voltage of this type of converter, a feedback type of control can be used. In general this control loop consists of two blocks, a difference block and a voltage-to-time converter. This is shown in Fig. 5.10.

This feedback type of control potentially introduces a number of additional poles. This is demonstrated in Fig. 5.11a, where the open-loop gain H and phase of the control loop are depicted. The most important pole is the Converter Pole and is located at a variable frequency due to the operation-point dependency of this pole. The second pole is the Control Circuit Pole and is located at a fixed frequency.

[5] The analysis presented in Sect. 5.2 gives enough information to build a control loop based on conventional PWM control used in inductive converter design (Wens and Steyaert 2011b).

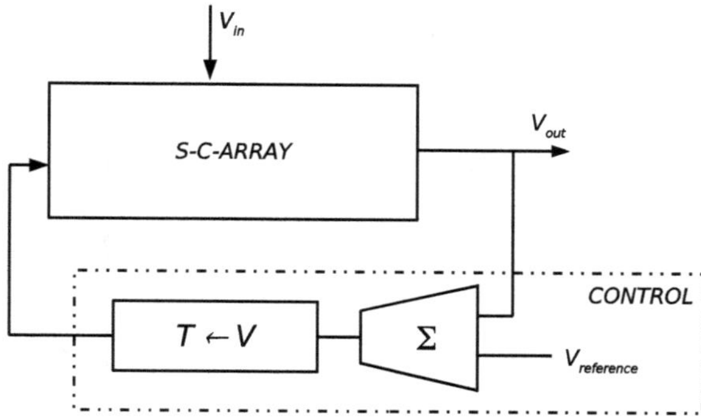

Fig. 5.10 Setup of a continuous control scheme for a capacitive DC–DC converter

These poles can reduce the loop's phase margin, which results in system instability. A first approach is to introduce a dominant compensation pole in the loop, as is demonstrated in Fig. 5.11b. If this intentional loop pole is dominant in the loop response, large enough phase margin is consolidated. But pole locations vary with the load conditions. Therefore for every specific converter implementation, worst-case pole locations are extracted to ensure stability for every operation point.

A more effective approach is to leave the pole from the power stage for what it is and to concentrate on the poles added in the control loop. This approach is much more effective since the pole in the power stage is operation-point dependent and the poles in the control loop have a fixed value independent from the converter's operation point. Thus by adding a zero, shown in Fig. 5.12b, the control loop poles are compensated in a robust way and the system is virtually reduced to a single-pole system and a high level of stability is observed.

The dynamic response of the DC–DC converter remains relatively slow and depends on the operation point of the converter. To omit this bottleneck, a discrete time–hysteretic-control method is proposed.

Discrete-Time Control

To develop a discrete-time hysteretic or comparator-based control technique, first the state-space nature is derived and graphically represented. In contrast to the conventional control techniques, no small-signal approximation or linearization is required.

State-Space Analysis Capacitive DC/DC-converters are Variable Structure Systems (VSS) that transfer charges from their input to their output by means of capacitors. In order to demonstrate the discrete-time control, a divide-by-two series-parallel converter is used as an example.

The divide-by-two series-parallel converter structure is shown in Fig. 5.2a. This topology consists of a charge-transferring capacitor (the flying capacitor C_{fly}), a buffer capacitor C_{out} and four switches. The DC–DC converter operates in two states. In state ϕ_1, shown in Fig. 5.2a, the flying capacitor C_{fly} is connected between

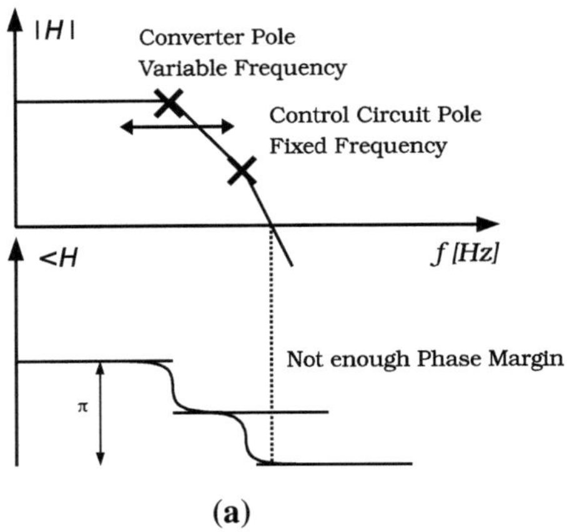

Fig. 5.11 **a** Simplified open-loop transfer function of a capacitive DC–DC converter, demonstrates small phase margin; **b** compensation of the open-loop transfer function by means of introducing a dominant pole results in conditional improved phase margin

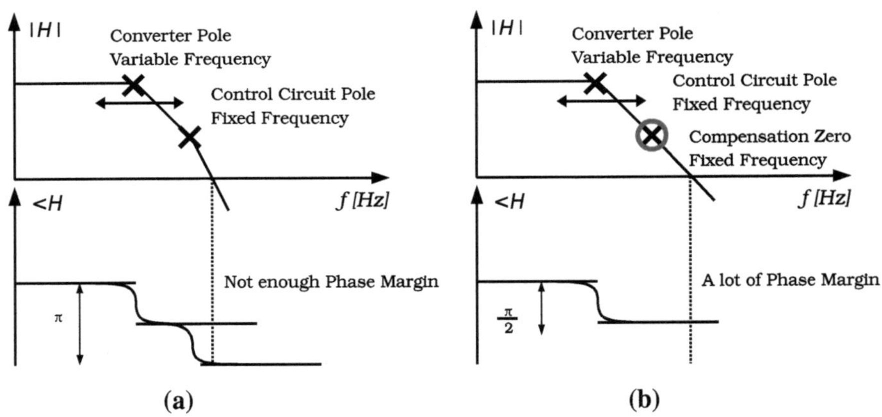

Fig. 5.12 **a** The simplified open loop transfer function of a capacitive DC–DC converter, demonstrates small phase margin; **b** compensation of the open loop transfer function by means of introducing a zero which eliminates the effect of the control circuit pole

input and output. During the second state ϕ_2, C_{fly} is connected between the output node and ground (Fig. 5.2b). By switching between these configurations charge is transferred from the input to the output.

The resistance R_s represents the equivalent aggregate resistance of the switches and the parasitic resistance of the capacitor. This VSS has a two-state nature and each state is described by a set of differential equations (Eqs. 5.26–5.29).

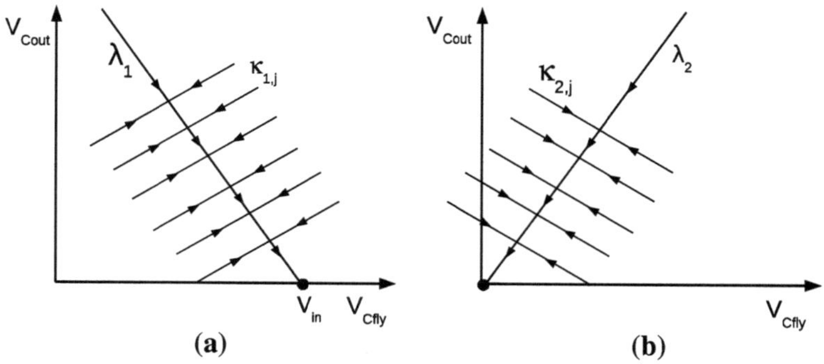

Fig. 5.13 The converter's state-space trajectories of state 1 in (**a**) and state 2 in (**b**)

State ϕ 1:

$$C_{fly}\frac{dV_{C_{fly}}}{dt} - C_{out}\frac{dV_{C_{out}}}{dt} = \frac{V_{C_{out}}}{R_L} \tag{5.26}$$

$$V_{C_{fly}} = V_{in} - V_{C_{out}} - R_s C_{fly}\frac{dV_{C_{fly}}}{dt} \tag{5.27}$$

State ϕ 2:

$$-C_{fly}\frac{dV_{C_{fly}}}{dt} - C_{out}\frac{dV_{C_{out}}}{dt} = \frac{V_{C_{out}}}{R_L} \tag{5.28}$$

$$V_{C_{fly}} = V_{C_{out}} - R_s C_{fly}\frac{dV_{C_{fly}}}{dt} \tag{5.29}$$

Continuous Dynamic systems are modeled by means of a State Space Model. The State Variables $V_{C_{fly}}$ and $V_{C_{out}}$ fully determine the operation of the converter in a unique way and enable analysis of the system dynamics. The system differential equations are decoupled in order to comply with the general form of the state matrix in Eqs. 5.30–5.31.

$$\dot{\vec{X}} = A\vec{X} + B\vec{U} \tag{5.30}$$

$$\vec{X} = \left(V_{C_{fly}} V_{out}\right)^T \tag{5.31}$$

Based upon the following equations the behavior in State Space can be calculated. For each configuration the vector field is represented in respectively, Fig. 5.13a, b.

State ϕ 1:

$$\frac{dV_{C_{fly}}}{dt} = \frac{V_{in} - V_{C_{out}} - V_{C_{fly}}}{R_s C_{fly}} \tag{5.32}$$

$$\frac{dV_{C_{out}}}{dt} = -\frac{V_{C_{fly}}}{R_s C_{out}} - \left(\frac{1}{R_s C_{out}} + \frac{1}{R_L C_{out}}\right)V_{C_{out}} + \frac{V_{in}}{R_s C_{out}} \tag{5.33}$$

State ϕ 2:

$$\frac{dV_{C_{fly}}}{dt} = \frac{V_{C_{out}} - V_{C_{fly}}}{R_s C_{fly}} \tag{5.34}$$

$$\frac{dV_{C_{out}}}{dt} = \frac{V_{C_{fly}}}{R_s C_{out}} - \Big(\frac{1}{R_s C_{out}} + \frac{1}{R_L C_{out}}\Big) V_{C_{out}} \tag{5.35}$$

It can be observed that every state has a distinct equilibrium point that will be reached if no change of configuration occurs. One can distinct between two type of trajectories: the main trajectories λ_i and their branch trajectories $\kappa_{i,j}$. For each configuration the branch trajectories lead to the main trajectories. For configuration 2 $(0, 0)$ is an equilibrium point and for configuration 1 $(V_{in}, 0)$ is an equilibrium point. By eliminating time from Eqs. 5.32–5.36 and Eqs.5.37–5.38, the slope of the trajectories can be found:

$$\frac{dV_{C_{out,1}}}{dV_{C_{fly},1}} = K_C\Big(1 - \frac{K_R V_{C_{out}}}{V_{in} - V_{C_{out}} - V_{C_{fly}}}\Big) \tag{5.36}$$

$$\frac{dV_{C_{out,2}}}{dV_{C_{fly},2}} = -K_C\Big(1 - \frac{K_R V_{C_{out}}}{V_{C_{out}} - V_{C_{fly}}}\Big) \tag{5.37}$$

$$with$$

$$K_C = \frac{C_{fly}}{C_{out}} \tag{5.38}$$

$$K_R = \frac{R_s}{R_L}$$

The main trajectory of configuration 1 is given in Eq. 5.39, the one of configuration 2 in Eq. 5.40. They intersect at $(\frac{V_{in}}{2}, \frac{V_{in}}{2})$. Equations 5.37 and 5.38 show that for a state on a considerable distance ϵ from the main trajectories, the slope is respectively K_C and $-K_C$. K_R partially determines ϵ. Thus it is clear that both K_C and K_R play a dominant role in the systems behavior.

$$\lambda_1 : V_{C_{out}} = V_{in} - V_{C_{fly}} \tag{5.39}$$

$$\lambda_2 : V_{C_{out}} = V_{C_{fly}} \tag{5.40}$$

By merging both vector fields a first impression is found in the interaction of both configurations on the system's trajectories (Fig. 5.14a) and the upwards evolution of the converter states in Fig. 5.14b.

Control Method. Control is performed by introducing a constant switching manifold ψ. This manifold is defined in Eq. 5.41. In conventional sliding mode control, the equilibrium points are separated by the switching manifold, this is not true for the switching manifold ψ. This is countered by the fact that for both configurations the branch trajectories direct the system towards ψ if $\psi < 0$. Both main trajectories tend however towards $V_{C_{out}} = 0$ if $\psi > 0$, but this is unwanted. Therefore a twofold

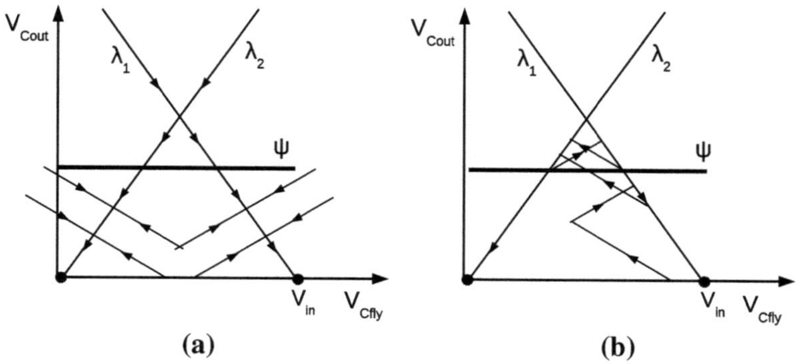

Fig. 5.14 **a** The combined trajectories of both conversion states; **b** the trajectory path of a capacitive DC–DC converter starting-up and evolving towards steady-state

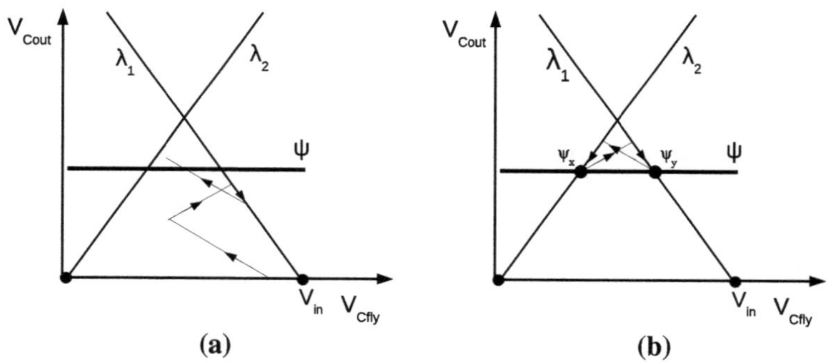

Fig. 5.15 **a** The start-up trajectory of a voltage divider; **b** the steady-state trajectory of a voltage divider: the so-called bow tie

switch nature is introduced. This is described in the two next paragraphs.

$$\psi : V_{Control} - V_{C_{out}} = 0 \tag{5.41}$$

For $V_{Control} - V_{C_{out}} > 0$ a fixed switching frequency is enforced. This results in a periodic reconfiguration of the DC/DC-converter, and the system evolves staircase-like towards ψ (shown in Fig. 5.15a). The question remains: what switching frequency should be enforced? It can readily be observed that if the switching frequency is relatively low and the state is not on the main trajectory, that the system first will evolve upwards according to the branch trajectory and next encounters a main trajectory and evolve downwards. The maximum switching frequency is therefore chosen so that the switching period corresponds to the time to fully charge the flying capacitor. If the switching period is too low. the system will never start up properly or will hover below the switching manifold without ever reaching it.

Once the switching manifold ψ is reached, the converter will only change configuration when the switching manifold is encountered. This occurs either in points ψ_x

and ψ_y. When the system resides in a state for which $V_{Control} - V_{C_{out}} < 0$ the only states to encounter the manifold are crossing points between the main trajectories and ψ. Thus reconfiguration occurs ideally only in these crossing points and since for each configuration there is a single branch trajectory, a unique path will be followed towards the other crossing point. Therefore ideally a bow tie-like trajectory arises (shown in Fig. 5.15b).

5.4 Implementations

In the previous section the most important frequency-domain analysis-techniques have been demonstrated and a number of control techniques have been described. This section focuses on how these control techniques are implemented by means of electronic circuits.

5.4.1 Lead Compensation

Capacitive converters can be controlled by translating an output-voltage error in a switching frequency error and consequently correcting the frequency error as is demonstrated in Fig. 5.16. The use of this technique imposes a frequency compensation scheme to obtain an stable control loop. Section 5.3, which demonstrates frequency-domain lead compensation by means of a zero, can be used to adopt frequency modulation as a capacitive converter's control method. Two important building blocks are required to accomplish this. The first one is the difference block, which calculates the control error and the second one is the voltage-to-time converter.

The difference block is implemented by means of a difference amplifier. The Operational Transconductance Amplifier (OTA) is one of the possible types of signal amplifiers used for this purpose. The choice of amplifier depends on the required control bandwidth, gain G_{amp} and the power consumption of the amplifier, given the converter characteristics. The loop gain of the converter system determines the static offset of the control loop. And the amplifier's bandwidth determines the frequency-domain pole locations and hence influences the stability of the closed-loop system.

The choice of OTA topology is determined by the required Gain-Bandwidth product (GBW), the input/output-voltage range and the available power budget. For a single-topology capacitive converter—operating in steady-state—the output voltage of the converter is committed to small voltage excursions and thus a full-range input is not required. Considering the input of the VCO at the output of the OTA, no full swing output is required either. The choice between PMOS- or NMOS-differential pair is determined on grounds of the relative position of the steady-state voltage between the OTA's supply rails. In many cases the control loop has separate supply rails with a lower supply voltage to achieve low power consumption. Hence the sensing of the output voltage is performed by means of a resistive

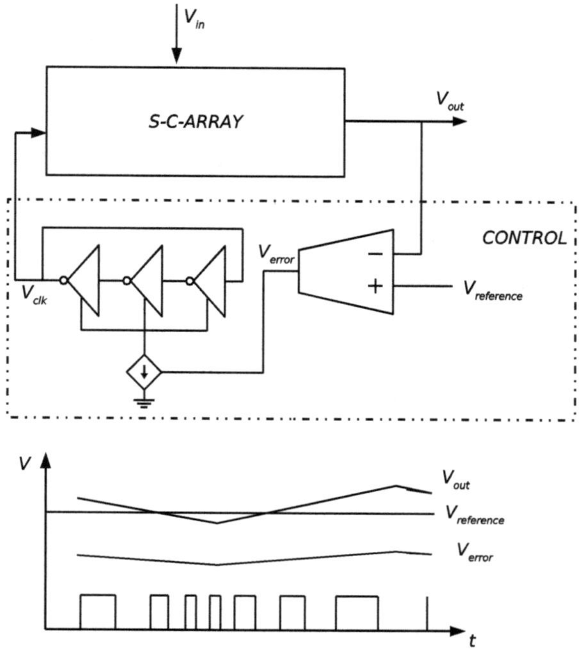

Fig. 5.16 Symbolic representation of a continuous-time-controlled capacitive DC–DC converter

divider. An example of a symmetrical OTA (single ended) is given in Fig. 5.17. It is clear that the input range of the NMOS-input topology is between V_{loop} and $V_{ds,currentsource} + V_{th,inputpair}$, while for the PMOS-input topology the input range is between V_{loop} and $V_{ds,currentsource} - V_{th,inputpair}$. The output voltage range of the OTA is cropped by the drain-source voltages across the output stage devices. The performance of this simple structure is boosted by adding current bleeding at the inner current mirrors (increasing power consumption) or adding cascode device in the Class-AB output stage (reducing the output voltage swing). If large input voltage swing is required, a folded cascode with complementary input pair can be selected (Sansen 2006).

As a voltage-to-time converter, the voltage-controlled oscillator (VCO) is used. For monolithic integration of a low-power oscillator the relaxation-type oscillator is preferred. This type provides an inductor-less solution and is therefore favored although the output waveform demonstrates a lot of distortion and is vulnerable to phase noise. Since the latter aspects are irrelevant for the application in a DC–DC converter control loop, the relaxation oscillator is an obvious choice. The current-starved ring oscillator is represented in Fig. 5.18. It comprises a ring oscillator that is fed by an array of current sources, both at the ground side as at the supply side, so that the charge and discharge rate of the inverter stages is controllable. The current sources are steered by a biasing circuit derived from a single-input voltage-controllable current source. The symmetrical biasing is required to obtain symmetrical rise and fall times and to ensure a 50 % duty cycle of the signal.

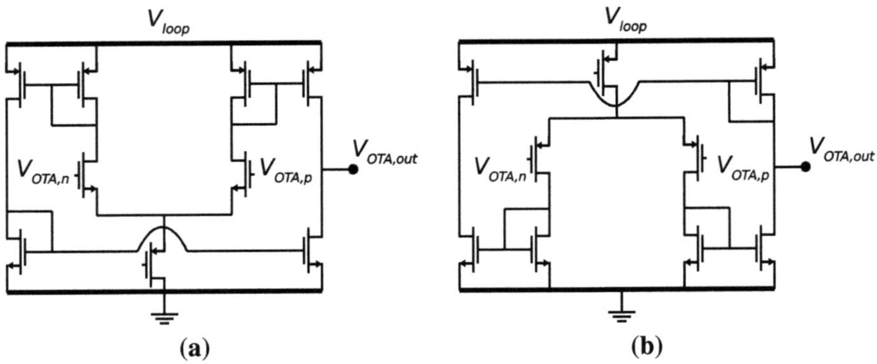

Fig. 5.17 Single ended symmetrical OTA. **a** NMOS-input differential pair and **b** PMOS-input

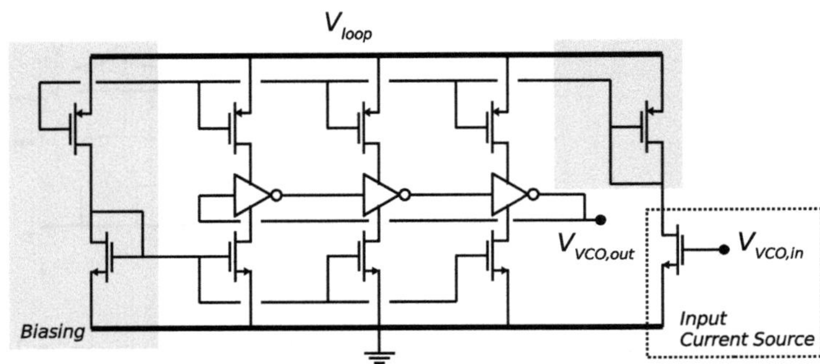

Fig. 5.18 Current-starved ring oscillator, frequency of the output voltage square wave is determined of the voltage applied to the input of the oscillator

In Fig. 5.16 a first straightforward frequency-modulated control loop is shown. The OTA generates an error voltage by comparing the output voltage with the reference voltage. The reference voltage corresponds to the desired output voltage. This error voltage steers the VCO's current source and determines the switching frequency's period (T). Compensation is obtained by inserting a lead circuit in the feedback loop. The lead circuit is demonstrated in Fig. 5.19. It comprises a resistor and a capacitor and can be incorporated in the resistance ladder used for level shifting the output voltage of the DC–DC converter within the input range of the OTA's differential pair. By positioning the zero[6] at the opamp's dominant pole and shifting the non-dominant poles beyond the switching frequency a stable control loop is obtained. Take into account that the resistance ladder simultaneously introduces a pole, but this pole is also shifted to a frequency beyond the switching frequency so that the relevant frequency response is not altered.

[6] The pole is located at a frequency of $\frac{1}{2\pi R_{div1} C_{lead}}$

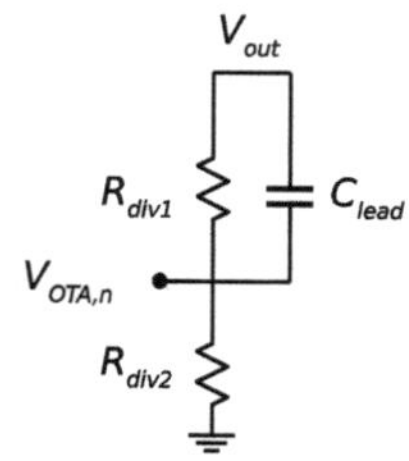

Fig. 5.19 Frequency lead circuit, it provides a gain equal to $\frac{R_{div2}}{R_{div1}+R_{div2}}$ at 'low' frequencies and a gain of one at 'high' frequencies

5.4.2 Lead Compensation for Multi-Phase Converters

In Chap. 4 multi-phase interleaving is proposed as an adequate technique to reduce the converter's current pulses and decrease the levels of noise at the input and the output node. Instead of a single clock signal, N_{MP} clock signals are required. Each of these signals have the same period T but are delayed $\frac{T}{N_{MP}}$ with respect to each other. This can be achieved by using an N_{MP}-tap voltage-controlled oscillator and using each of the VCO's phases to address one of the N_{MP} converter cores. This is demonstrated in Fig. 5.20. Again a current-starved VCO is used for implementing this functionality, but due to the nature of the VCO only an odd-numbered number of phases can be driven. For use in even-numbered interleaved converters differential-type VCO's must be used.

Where the previous implementation results in a quite compact solution, it becomes very hard to test the converter in case the VCO does not function as expected. Therefore the implementation presented in Fig. 5.21 provides an alternative solution. Instead of using a clocking signal from each separate VCO-tap, a single signal of the VCO is used and from this signal the separate clocking signals for the converter phases are derived.

This has the advantage that for testing only a single signal has to be overruled, instead of N_{MP}. The VCO runs at an N_{MP} times higher frequency. Depending on the power consumption and the need for testability of the application one or the other solution is opted for.

5.4.3 Hysteretic Discrete-Time Control

In the previous section the state-space behavior of a capacitive converter has been analyzed. The most important observation is that the operational state-space is confined to a surface bound by the converter's nature at one side. For each arbitrary state the converter operates as such that the converter's state migrates to the area between the main trajectories. So the converter is bounded by nature and only a single additional boundary is necessary to control the converter's state.

In Fig. 5.22 the hysteretic discrete-time controller is shown in principle. It operates as follows. The comparator detects when the output voltage crosses the control

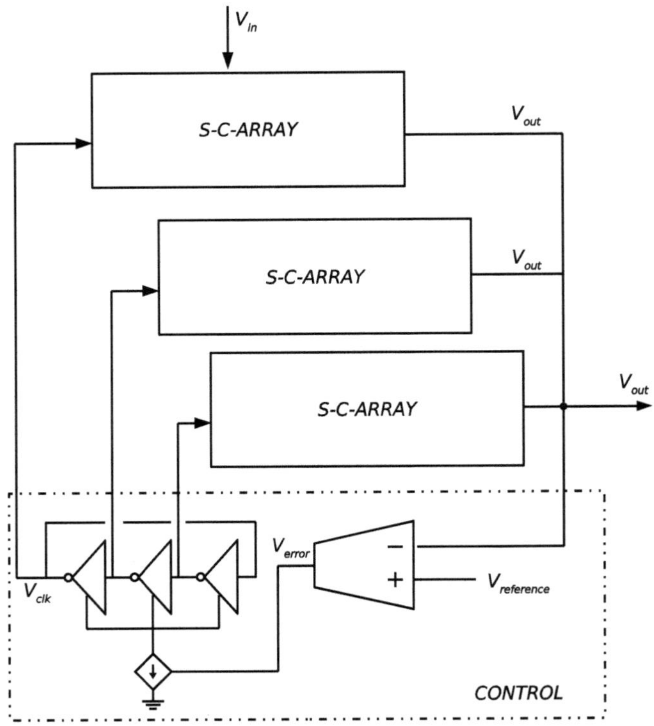

Fig. 5.20 Using a continuous control scheme to apply control to a multi-phase capacitive converter

boundary $V_{reference} - V_{C_{out}} = 0$, for an external reference voltage at point (a). By adding an edge-triggered latch the sample clock frequency V_{clk} is eliminated such that only a rising edge is associated with a charge transfer. If the latch is omitted the falling edge at point (b) which is an artifact of the comparator type and not a trigger for a boundary crossing, will trigger a charge transfer. This charge transfer is unwanted. V_{latch} corresponds to the actual switching frequency of the converter and initiates thus a charge transfer to the output.

5.4.4 Multi-Phase Hysteretic Discrete-Time Control

Since the multi-phase interleaving approach becomes the dominant technique to reduce converter noise and increase power density in integrated converters, unification of the multi-phase technique with the hysteretic approach is of high importance. This way high bandwidth at low power cost is combined with ripple reduction and an increase in power density.

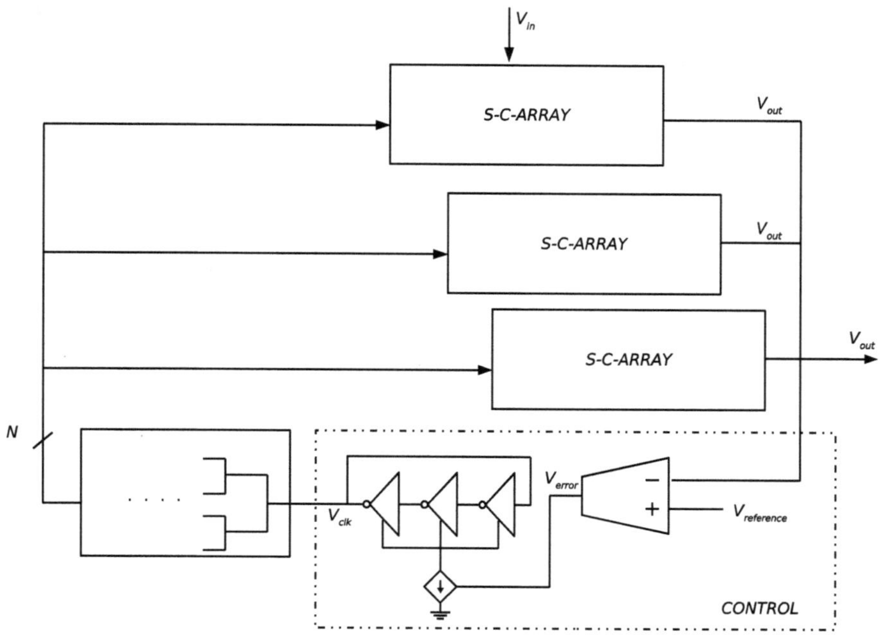

Fig. 5.21 Using a continuous control scheme to apply control to a multi-phase capacitive converter

Multi-Loop Multi-Phase

Essentially multi-phase interleaving comprises smearing out the charge transfer in time. The hysteretic controller is implemented by means of a clocked comparator. The latter turns the single-phase controller into a synchronous circuit. In order to prevent simultaneous switching of multiple paralleled capacitive converters, each converter is clocked by a different signal and these signals are phase-shifted with respect to each other.

In practice this comes down to a separate hysteretic control loop for each converter core and these are fed by signals coming from a single clock but with a shift in phase. This is shown in Fig. 5.23. The converters can be observed as N_{MP} single-phase converters connected to the same output capacitor. For every individual converter the charge transfer of the other converters is nothing else but an external disturbance. Reaction on this is not different from the reaction on a change in the load condition.

The drawbacks of this approach are linear (factor N_{MP}) increase in control circuitry and thus silicon area consumed by the circuitry and the matching requirements of the comparators. Moreover this technique does not guarantee equal current sharing between the converters especially not in case there is a significant difference in comparator offset. Offset tuning of a set of comparators seems unrealistic for this kind of applications and therefore a single-loop approach can be preferred.

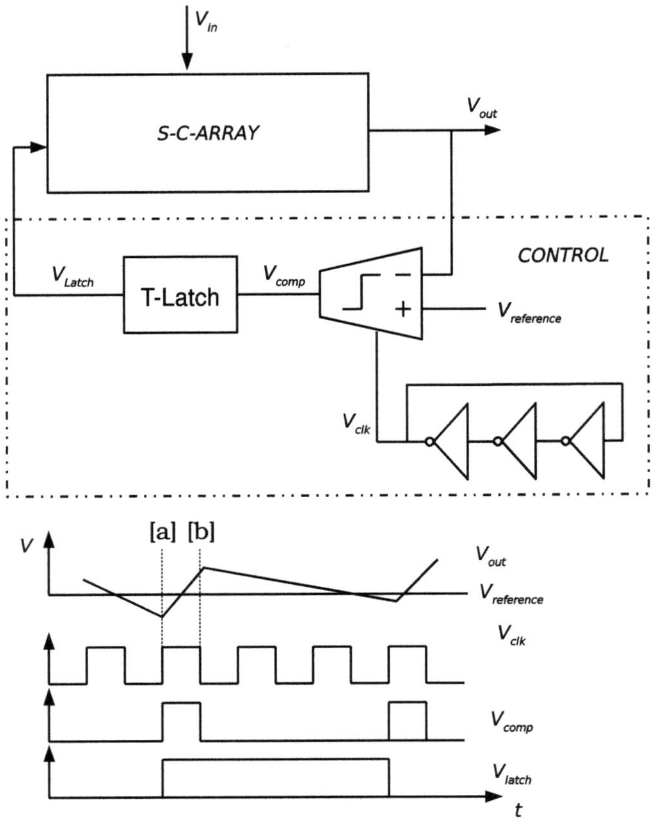

Fig. 5.22 Single phase hysteretic controller used for PFM control of a capacitive DC–DC converter

Single-Loop Multi-Phase Handover

The weakest link of the multi-loop multi-phase control loop is the use of multiple comparators to define a single control boundary. Therefore an alternative control technique is developed, omitting all but one comparator.

The single loop phase-handover control for capacitive converters is shown in Fig. 5.24. Each of the converter cores is connected to a series of cross coupled flip flops: the phase-handover-block. This block has N complementary outputs (thus 2N in total) from which only one pair of signals is toggled each period of the block's input signal.

Where the non-dominant pole compensation technique uses the VCO to guarantee sequential activation of the cores, the phase-handover block plays this role in this hysteretic incarnation. This has the important advantage with respect to the multi-loop approach that irrespective of the load, the cores have an equal degree of activity. This impedes overstressing part of the converter, which is particularly desirable for

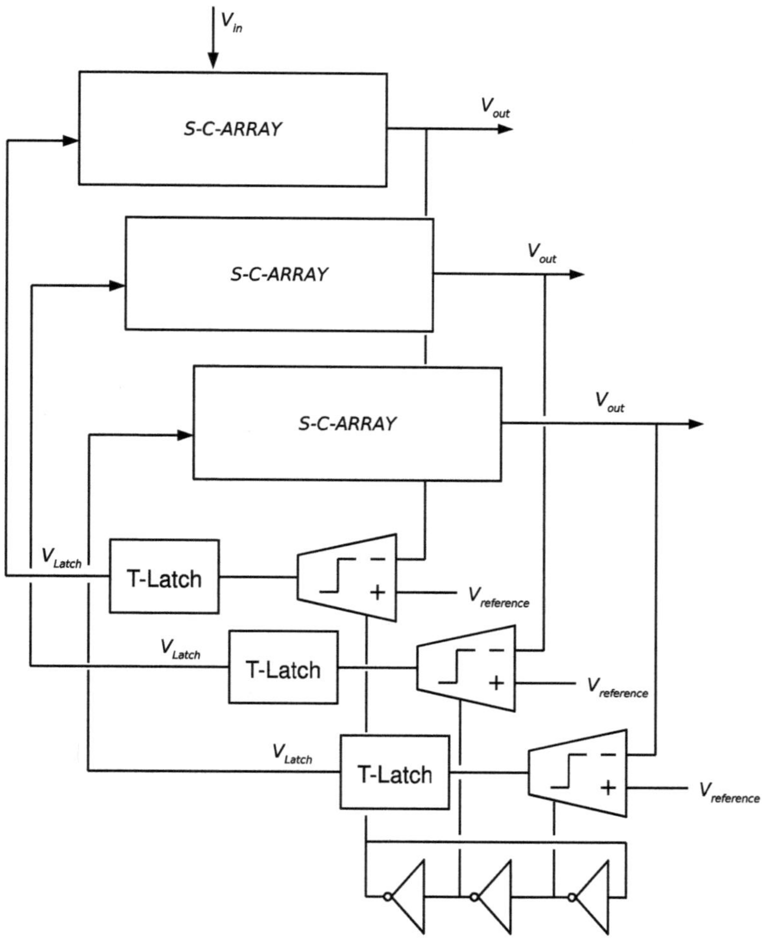

Fig. 5.23 Multi-loop multi-phase hysteretic controller

high-power converters due to the thermal aspects of this overstress. If the hysteretic controller is connected to the clock of the phase-handover block, each output voltage boundary violation is handled by a different core. This results in a perfect current-sharing phase distribution. Moreover only a single comparator is included in the loop, which eliminates the mismatch issue discussed in the previous paragraph. In fact the T-Latch can be eliminated when the phase-handover block takes over the falling-edge filtering. The sampling frequency of the loop is higher but this is countered by the reduction in loop circuitry compared to the multi-loop approach.

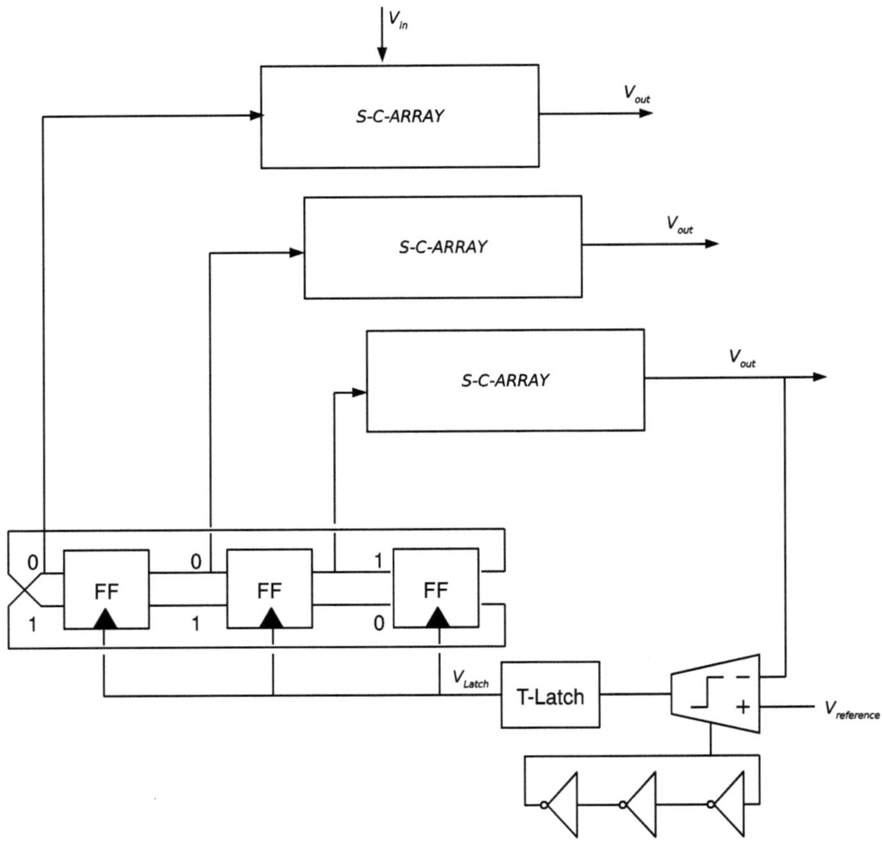

Fig. 5.24 Single-loop multi-phase hysteretic controller

5.5 Conclusions

Control of capacitive DC–DC converters is one of the most crucial properties to guarantee successful adoption of monolithic DC–DC converters in a power-management set-up. Essentially topology reconfiguration and PFM control are excellent control methods to be applied in control of monolithic capacitive DC–DC converters. Topology control addresses the VCR and topology relationship. PFM control is used for load regulation and is especially beneficial for maintaining a flat conversion efficiency curve over a broad output power range.

By combining hysteretic control with multi-phase interleaving, as is demonstrated first in Van Breussegem and Steyaert (2010a), a versatile low-power control method is conceived. First of all since multi-phase interleaving has proven its merit as a powerful technique to reduce ripple and increase power density for fully integrated DC–DC converters. Secondly since hysteretic control demonstrates a high-bandwidth control technique, relying on an all-digital circuit implementation. Combining this with a multi-topology converter, shown in Van Breussegem and Steyaert (2010b), results in an extremely versatile power-management interface. This versatility manifests itself in a broad input-output voltage range, high potential efficiency, a high control bandwidth for as well line regulation as load regulation and conservation of the efficiency over a broad output power range.

Chapter 6
Monolithic Integration of DC–DC Converters in CMOS

The previous chapters approach the capacitive DC–DC converter in a generic manner. Essentially a capacitive DC–DC converter is a structure of both active and passive components, the switches and the capacitors, which are structured according to a specific topology described in Chap. 2 and reconfigured based on a control method described in Chap. 5. These chapters take a number of imperfections into account. As is illustrated in Chap. 3 these imperfections are the dominant restrictions when designing a circuit or in this case a high-performance capacitive DC–DC converter.

The aim of the forthcoming chapter is to match the imperfections discussed in the previous chapters with the actual technology used for building the integrated circuits. Complementary Metal Oxide Semiconductor-CMOS in short—is the dominant technology used for integration of high-performance digital and analog circuits. It is the foremost important technology for implementation of the state-of-the-art mass-production microcontrollers, microprocessors and many other electronic systems used in everyday life. The following sections give a short overview of the technology and the important features when used for implementation of capacitive DC–DC converters.

6.1 Technology Framework

CMOS technology comprises a number of materials and techniques to build electronic circuits in a single substrate. The substrate is a p-doped silicon slice, shown in Fig. 6.1a. By applying a number of processing steps to this p-doped substrate, devices are formed in the substrate. In Fig. 6.1b the cross-section of a CMOS-transistor is depicted. This transistor can either been used as a switch or as a voltage-dependent current source. By means of the same techniques passive devices are conceived: integrated resistors and integrated capacitors. These devices are then connected in Fig. 6.1c by means of metallic or poly-silicon wires. By connecting these devices, circuits are formed. In their turn these circuits are used in large electronic systems.

T. Van Breussegem and M. Steyaert, *CMOS Integrated Capacitive DC–DC Converters*, Analog Circuits and Signal Processing, DOI: 10.1007/978-1-4614-4280-6_6,

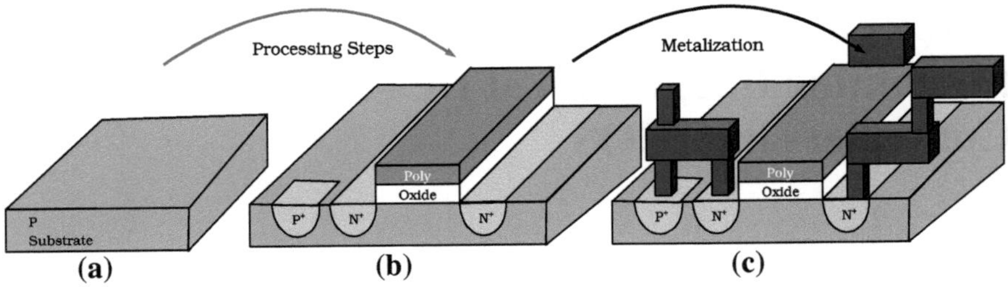

Fig. 6.1 **a** Construction of integrated circuits from a slab of doped silicon. **b** First a number of processing steps are executed so that solid-state devices are formed. **c** The devices are connected by metal or poly-silicon wires to form circuits

The CMOS process is a planar process, therefore the most important dimension is the area. This area is strongly related to the cost of the circuit, especially important for mass-production designs.

6.2 Solid-State Switches

The most important active devices are the transistors. They introduce electrical controllability, gain and switching properties in the circuits. First the fundamental behavior of a transistor is explained and next two techniques are discussed to overcome the voltage restrictions of the state-of-the-art transistors.

6.2.1 Operation Regions

The transistor is a four-terminal circuit element. In Fig. 6.2 an enhancement N-type MOS transistor is depicted. It consists of two N^+ doped regions: the drain and the source region. In between both regions an oxide layer with a poly-silicon layer on top of it forms the third terminal: the gate of the transistor. The fourth terminal is formed by the P-type substrate in which the source and drain are embedded: the bulk of the transistor. Based on the voltages across these terminals the transistor resides in one of the following operation regions: saturation, linear, sub-threshold or cut-off. The important voltages are the voltage between the source and drain terminal: V_{DS}, the voltage between gate and source: V_{GS} and the threshold voltage V_t. The latter is a characteristic of the transistor instead of a measurable voltage. For sake of clarity, the transistor is looked at from a functional point of view instead of a device physical point of view.

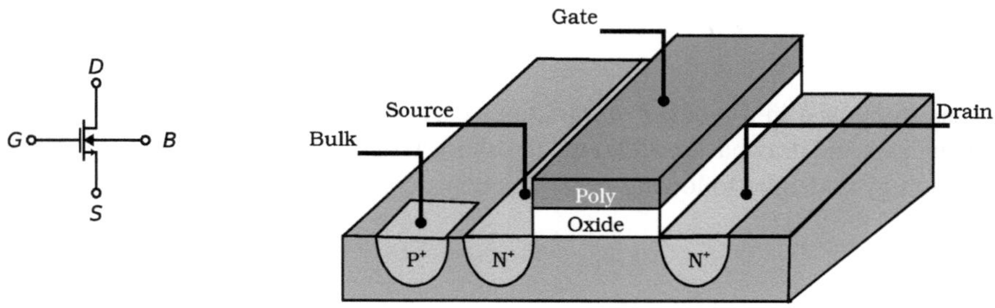

Fig. 6.2 Lateral section of a NMOS-type transistor. The four terminals are indicated and the type of diffusion is typical for an NMOS enhancement-type transistor

Saturation Region

The transistor resides in the so-called saturation region if: $(V_{GS} - V_t) < V_{DS}$. We assume that $V_{GS} > 0$, $V_{BS} = 0$ and $V_{DS} > 0$. A current I_{DS} flows through the transistor from the drain terminal to the source terminal.

Then:

$$I_{DS} = \frac{\mu_n C_{ox}}{2} \frac{W}{L} (V_{GS} - V_t)^2 (1 + \lambda V_{DS}) \tag{6.1}$$

with, μ_n represents the electron mobility, W and L respectively the width and length of the transistor's channel and λ a technology constant. When neglecting λ it is clear that the current is a quadratic function of the overdrive voltage $V_{od} = (V_{GS} - V_t)$. Thus the transistor can be used as a voltage-dependent current source as long as the saturation-boundary conditions are valid. In DSM technologies λ can no longer be neglected and an effect called channel-length modulation comes in the picture: the additional dependency of the current on the drain-source voltage V_{DS}.

Linear Region

The transistor resides in the linear region if: $(V_{GS} - V_t) \geq V_{DS}$. A current I_{DS} flows through the transistor from the drain terminal to the source terminal:

$$I_{DS} = \mu_n C_{ox} \frac{W}{L} \left((V_{GS} - V_t) V_{DS} - \frac{V_{DS}^2}{2} \right) \tag{6.2}$$

Taken that the second term is negligible, the transistor behaves as a voltage dependent resistor. In extremis by applying large gate-voltage variations it can be used as a switch with a non-zero on-resistance. The on-resistance can be made arbitrarily small by keeping the length of the transistor as small as possible—determined by the process minimum feature size—and make the width of the transistor as large as required.

6.2.2 Transistor Flavors

Transistors exist in a number of variations each with their specific characteristics, advantages and drawbacks. Although the characteristics are tightly connected to the technology node, the following general conclusions can be made.

NMOS Type

The NMOS type transistor is the most simple device in the CMOS stack and requires the least process steps as is shown in Fig. 6.3a. This device is used mostly for low-side switches in buck converters and in capacitive converters for switches that have a large headroom from the source up to the positive rail. The main drawback of this device is the lack of isolation of the channel. Moreover the doped source and drain regions are only separated by a diode from the substrate. To operate properly this diode must remain reverse-biased at all time. The lack of isolation from the substrate renders it impossible to use the NMOS type transistor as a flying capacitor. On the other hand the NMOS transistor is intensely used as output buffer capacitor, since isolation is not necessary in this case. This will be discussed in the section on the passive devices.

PMOS Type

The PMOS-type transistor is the complementary device of the NMOS and requires an additional processing step: creation of an N-doped well before doping the drain and source regions. It also requires and additional terminal to connect the bulk (N-Well), demonstrated in Fig. 6.3b. The mobility of the carriers is lower in PMOS devices than in NMOS devices, which results in a higher channel resistance. As a consequence, given an equal overdrive voltage, the transistor is sized up to achieve an equally low on-resistance as an NMOS type transistor would.

Tripple-Well NMOS Type

By adding two wells: a P-doped Well and a Deep N-Well, the NMOS device's bulk is isolated from the substrate as is depicted in Fig. 6.3c. By doing this, the bulk voltage of the NMOS transistor is not necessarily at the ground voltage. This way, the threshold voltage can be reduced with respect to a non-zero V_{BS}. Reduction of the threshold voltage improves the conductivity of the switch since it increases the overdrive voltage V_{od}.

Additionally it isolates the lower plate of the MOS capacitor[1] from the substrate, so that the triple-well NMOS transistor can be used to implement flying capacitors.

[1] Discussed further in the next sections.

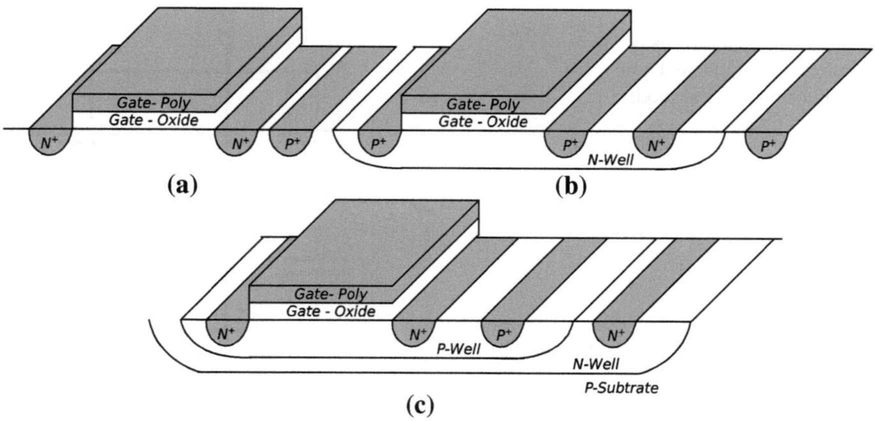

Fig. 6.3 **a** NMOS doping profile. **b** PMOS doping profile. **c** Triple well NMOS doping profile

High-Speed Type

The high-speed transistor is the regular transistor type which is aiming for high propagation speeds in digital circuits. Especially thanks to the low threshold voltage this can be achieved. But this goes at the cost of two heavy drawbacks. First the low threshold voltage results in larger leakage current, when the transistors—used as switches—are in the off state. Secondly the very thin oxide used in these devices demonstrates leakage currents through the gate of the transistor. When the transistors are used as capacitors, the so-called MOS capacitors, leakage through large capacitors is significant enough to affect DC–DC converter efficiency.

Low-Leakage Type

These devices have a slightly thicker oxide and a higher threshold voltage V_t but provide better isolation and negligible leakage current. The higher V_t and thicker oxide reduce the capacitance density.

Thick-Oxide Type

Thick-oxide devices are transistors with a larger minimum feature size (a larger minimum channel length L) than the standard devices and a thicker oxide. This makes them capable to deal with larger voltages and they are often used for building I/O-circuits or as high-voltage capacitors. The latter at the cost of capacitance density.

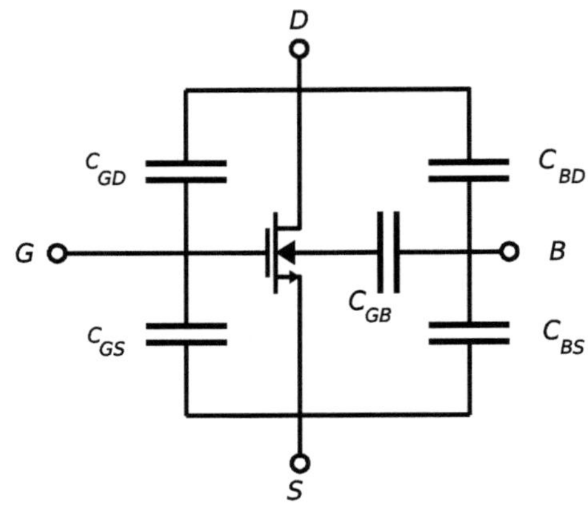

Fig. 6.4 Capacitance model of an NMOS-type transistor. In practice an equivalent gate capacitance is used to model the switching losses

6.2.3 Parasitic Elements

The parasitic elements of a component are the secondary characteristics of a component which are unwanted regarding the primary use of the component. Whether a certain characteristic is regarded as a parasitic characteristic or not is determined by the use of the component.

Parasitic Capacitance

Besides the gate capacitance of a transistor a number of additional capacitances play a role. Between every terminal of the transistor a capacitive coupling exists. Between gate and source/drain due to the overlap between the gate and drain and source regions. Between drain/source and bulk by means of the inversely polarized drain/source-bulk diodes. In Fig. 6.4 a model is demonstrated depicting the most influential parasitic capacitances of a DSM transistor. In order to facilitate insightful modeling the relevant capacitance is lumped in a single gate capacitance. This approach is validated by cross checking driving power consumption in the mathematical model and circuit simulation by means of Spice.

On-Resistance

The on-resistance of a transistor used in the triode - or linear region is:

$$R_{on} = \frac{V_{DS}}{I_{DS}} \tag{6.3}$$

By substitution of Eq. 6.2 in Eq. 6.3, the following expression for the switch's on-resistance is obtained:

$$R_{on} = \frac{1}{\mu_n C_{ox} \frac{W}{L}((V_{GS} - V_t) - \frac{V_{DS}}{2})} \tag{6.4}$$

Obviously a low R_{on} goes hand in hand with a high C_{ox} and a large overdrive ($V_{GS} >> V_t$). When a transistor is used as a switch, the primary characteristic is the infinite off-resistance of the transistor and the zero on-resistance. But operating a transistor as a switch requires changing the voltage V_{gate} between the gate and source terminal. The presence of the gate-capacitance C_{gate} implies that charge is required to change the voltage across the terminals. The process of charging and discharging the gate capacitor requires energy:

$$E_{charge} = C_{gate} V_{gate}^2 \tag{6.5}$$

Thus changing the state of the switch ($S = 0 \rightarrow S = 1 \rightarrow S = 0$) introduces an energy penalty and impedes us of making the size extremely wide so that the on-resistance is non-zero. In Chap. 3 it has been made clear that for an optimum implementation the different losses need to be in balance.

For the state-of-the-art DSM technologies the previously cited formula for on-resistance is not accurate enough. Two effects have to be added to the model to comply with the DSM technologies (BSI). First there is the effect of the gate-source voltage on the mobility. The latter is corrected by using an effective mobility μ_{eff} instead of the constant μ_n. The second additional effect is the Lightly-Doped Drain resistance (R_{LDD}). In order to reduce the hot-carrier effect, the source and drain area are lighter doped than the transistor's channel (Sheu et al. 1984). This introduces an additional series resistance. For most DSM technologies this additional term is inversely proportional to the transistor width. The resistance is approximated by:

$$R_{on} = \frac{1}{\mu_{eff} C_{ox} \frac{W}{L}((V_{GS} - V_t) - \frac{V_{DS}}{2})} + R_{LDD} \tag{6.6}$$

Additional Parasitic Elements

Next to the secondary characteristics, the ones included in the transistor model, a whole range of parasitic elements must be taken into account. In order to connect the transistor's terminals to the drivers and the capacitors a low-ohmic metal connection must be accomplished. Careful analysis is required when routing and connecting metals between building blocks or components. Especially in high-power-density converters, the metal interconnect forms the real bottleneck. Due to Design Rules of a technology the design freedom of certain components is limited and this might

Table 6.1 Maximum voltage of the native devices in some of the state-of-the-art technology nodes

Technology (μm)	V_{max} (V)
0.25	2.5
0.18	1.8
0.13	1.2
0.09	1.2
0.045	1
0.032	1

introduce unwanted and sometimes unexpected parasitic capacitance coupling with the substrate or parasitic resistance in the power plane of a DC–DC converter.

6.2.4 Dealing with Voltage Limitations

CMOS technologies are characterized by the maximum voltage between the gate and source terminal of the transistor. In Table 6.1 an overview is given of the maximum voltage rating along the most recent technology nodes of a standard CMOS process. It is observed that the maximum voltage decreases along with the minimal line width of the process node but saturates for the Deep Sub Micron nodes. The maximum voltage is dictated by the oxide break-down voltage and includes a reasonable margin to prevent performance degradation over longer time periods (Maricau and Gielen 2010). The maximum drain-source voltage is typically higher than the maximum gate-source voltage but the same voltage rating is applied by the process manufacturers.

Voltage break down is the most critical restriction when using a standard CMOS process for implementing an integrated DC–DC converter. Integrated DC–DC converters are providing an interface between a broad range of input voltages and a accurately regulated output voltage. These input voltages often exceed the maximum voltages of the state-of-the-art processes depicted in Table 6.1. In some technologies two oxide thicknesses are foreseen and high-voltage devices are processed. But if these are not foreseen or these devices still do not meet the specifications, circuit techniques are provided to bridge the gap. In the following paragraphs two techniques are elaborated: Switch Stacking and Voltage Domain Stacking.

Switch Stacking

The most popular technique to deal with higher voltages is by replacing each switch by a series-connected combination of two or more transistors. This is demonstrated in Fig. 6.5 for a buck converter. The buck converter, shown in Fig. 6.5a, is built by means of a sized-up inverter and an LC-filter. The inverter generates a square wave with variable duty cycle and the filter turns it into a DC voltage at the output. In the

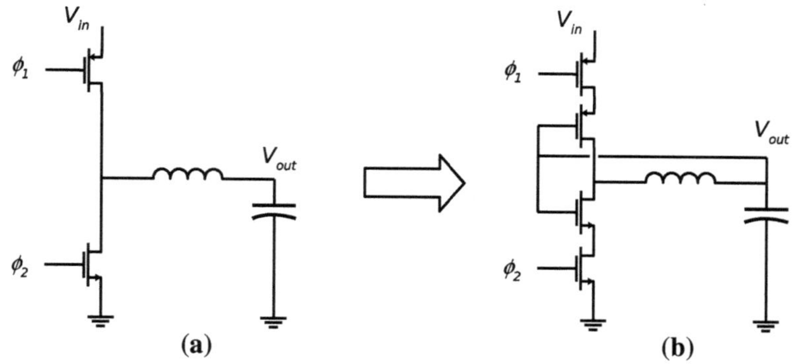

Fig. 6.5 Example of a conventional buck converter in (**a**) and a buck converter using the switch-stacking technique to overcome voltage headroom limitations in (**b**)

regular case, both the NMOS and PMOS transistor are exposed to the full swing input voltage. If the input voltage is higher than the break-down voltage of the transistors. an alternative topology, demonstrated in Fig. 6.5b, is used (Wens et al. 2008).

The voltage across the switch is divided over the two switches, so that each of the switches is exposed to only half of the total input voltage and none of the breakdown restrictions is violated. The use of this technique implies that level-shifted signals are generated to steer the high-side switch with signal Φ_1 and that both bulk and gate of the additional transistors are properly biased. For every topology a different biasing scheme is required (Wens et al. 2007, 2008). This technique is an effective technique but especially useful for DC–DC converters with a small number of switches, for example buck or boost-type converters. Capacitive DC–DC converters have typically a much larger number of switches and since switch stacking at least doubles the number of switches, it is less attractive the apply the technique.

Voltage Domain Stacking

A proper alternative for Switch Stacking is the Voltage Domain Stacking Technique. Instead of dividing the voltage over two devices, the circuit is constructed in such a way that due to the topology, none of the switches is exposed to a voltage higher than the maximum voltage rating. In practice some of the topologies are identified to support this principle by nature. For example the Ladder-Topology, shown in Fig. 6.6, is a topology which requires little adaptation to use this technique. In case of steady state and no load, the maximum drain-source voltage of the switches S_{1-6} is equal to a third of the input voltages. The switches are thus exposed to a certain fraction of the input range, this fraction is determined by the topology.

Proper operation implies that the swing at the gate terminal does not exceed the voltage ratings either. A bigger problem arises during the start-up procedure.

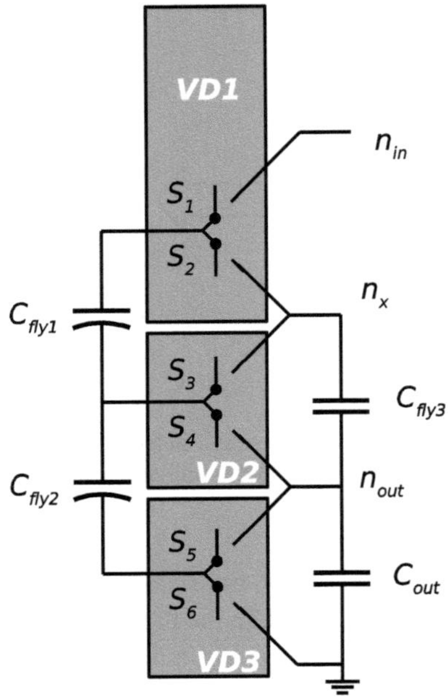

Fig. 6.6 Voltage domain stacking can be used in a capacitive DC–DC converter with a ladder topology

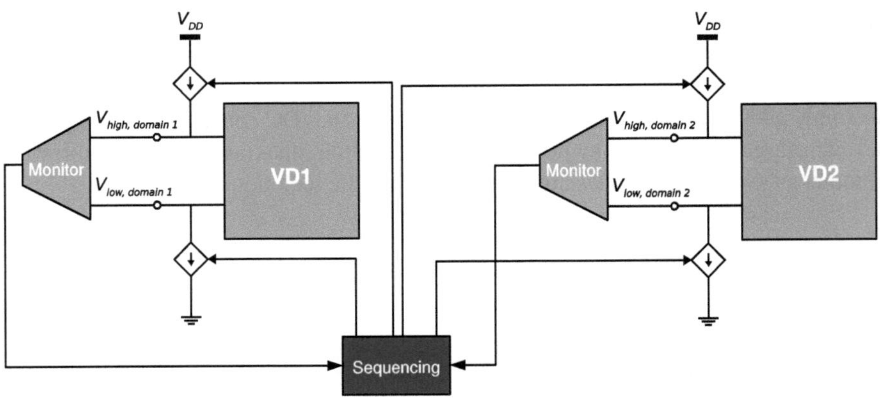

Fig. 6.7 By introducing multiple floating voltage domains, voltage-domain stacking is used to overcome the voltage limitations in low-voltage technologies. This requires voltage-domain monitoring and voltage-domain control

Additional circuitry for monitoring and start-up sequencing is required to protect the switches during this phase.

A technique is developed which monitors the voltage domains and corrects the voltage-domain rails so that the voltage-domain boundary conditions are ensured. In Fig. 6.7 this is schematically represented. In general each switch is enclosed by a single voltage domain. A monitor observes the voltage-domain rails and passes this information through to the sequencer. This sequences gathers data from all the voltage domains and processes it. Based on this the rails are adjusted so that the voltage domains are secured. In practice the voltage domains share rails and the decision making algorithm becomes less complex. If the topology supports multiple voltage domains (f.e. the Ladder Topology), the monitoring and sequencing are only addressed during start-up or heavy transients. This reduces the impact of this technique on the system's power consumption. In Chap. 7 a practical example of the voltage-domain technique is elaborated.

6.3 Passive Devices

6.3.1 Fundamentals

Capacitors are the key components of capacitive DC–DC converters. The flying capacitors provide the means of transferring the charge and the output buffer capacitor damps the current pulses so that a voltage with as little noise as possible is generated at the output. Integration of these capacitors puts heavy constraints on the design space of a capacitive converter. The main obstacles are the capacitance density and the parasitic capacitance of the integrated capacitors.

Capacitance Density The capacitance density $C_{\square}$ determines the cost of the capacitor. It is the main/useful amount of capacitance normalized by means of the area A occupied by the capacitor.

$$C_{\square} = \frac{C_{main}}{A} \tag{6.7}$$

The higher the density the higher amount of capacitance that can be integrated at the same cost. When a larger amount of capacitance is integrated, the switching frequency can be reduced. Then the switching losses are decreased and thus the overall converter efficiency improved.

Capacitance Quality The quality of the capacitor α also has a large impact on the converter's efficiency. In previous chapters, the quality factor is defined as the ratio of the parasitic capacitance C_{par} with respect to the amount of useful capacitance C_{main}:

$$\alpha = \frac{C_{par}}{C_{main}} \tag{6.8}$$

In case of a straight forward parallel-plate capacitor, it is the ratio of the parasitic capacitance between bottom plate and the substrate and the capacitance between bottom and top plate.

Capacitor's Resistance A third but not unimportant factor is the parasitic resistance of the capacitor. This resistance introduces additional power loss, increases the output

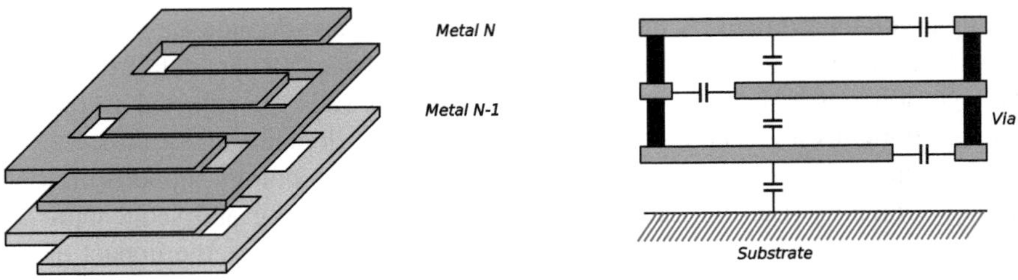

Fig. 6.8 Structure of Metal-Oxide-Metal Capacitors: Interdigited structures provide both vertical and lateral coupling

impedance which must be compensated by increasing the switching frequency again and introduces additional noise at the output.[2]

Therefore it is of high importance to analyze the available capacitors in the CMOS process and compare them based on their main characteristics. In this discussion the three dominant capacitor types are discussed: MOM capacitors, MIM capacitors and MOS capacitors.

6.3.2 Metal-Oxide-Metal Capacitors

Construction Metal-Oxide-Metal type of capacitors are formed by 2 to N layers from the CMOS metal stack (Samavati et al. 1998). Capacitance density is increased by using finger-type interdigited structures, shown in Fig. 6.8, which results in both lateral and vertical capacitive coupling. The capacitance density increases by increasing the number of metal layers and by reducing the inter-plate distance. But since the lower layers in the stack exhibit more capacitive coupling to the substrate, this also increases the capacitor's parasitic capacitance ratio (α) and reduces the quality.
Characteristics The parasitic series resistance of this structure is relatively low as a result of the full metal connections but degrades when finger length of the interdigited structure increases. Linearity of these devices is inherently good. Using this type of capacitor enables capacitance densities of about 0.2–1 fF per square μm (in state-of-the-art DSM CMOS) depending on how many metal layers are used and the pitch between these layers. The decreasing feature size in the DSM CMOS technologies provides excellent perspectives for improving the capacitance density of this type of capacitors.

[2] The latter is discussed in Chap. 4.

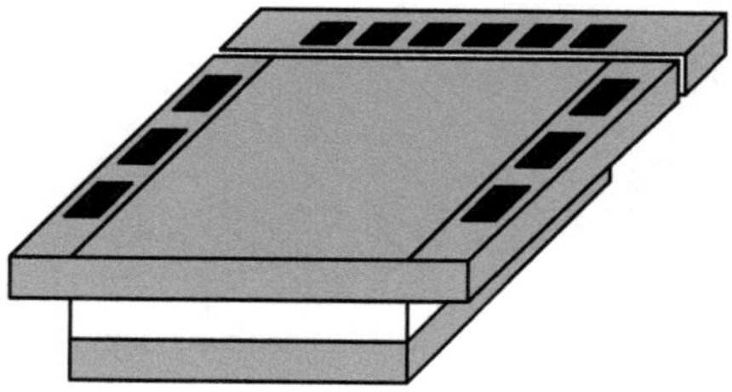

Fig. 6.9 Structure of a Metal-Insulator-Metal Capacitor, additional processing steps are required but enhance the capacitance density by using high-permittivity insulator and reduced plate distance

6.3.3 Metal-Insulator-Metal Capacitors

Construction MIM capacitors are conceived in the upper layers of the metal stack. It is a capacitor between a conventional metal layer (M_{N-1}) and an additional layer (M_x), as is shown in Fig. 6.9. This additional layer is placed between the M_N layer and the one above it (M_{N-1}) (Burghartz et al. 1997). To increase the capacitance density additional lithographic steps are added to the conventional digital CMOS process. First a step deposits a dielectric with a higher permittivity than the regular silicon oxide and afterwards an extra metal plate M_x is introduced at a smaller spacing than the regular metal pitch.

Characteristics Since this type of capacitor resides in the upper layers of the CMOS metal stack, the parasitic capacitive coupling between the capacitor plates and the substrate is rather small. The coupling is dominated by the coupling between the capacitors bottom plate and the substrate and can be reduced down to a few percentages ($\alpha = 1$–$2\,\%$) in Deep Submicron Technologies. The quality of this MIM capacitor can be improved by omitting dummy metal deposition in the underlying metal layers (as far as this is allowed by the design rules). The MIM capacitor's construction makes it highly linear and appropriate for use in topologies with small capacitor voltages. MIM capacitors obtain capacitance densities between 1 and 3 fF per μm^2.

6.3.4 Metal-Oxide-Semiconductor Capacitors

Construction Also MOS devices can be used as capacitors. The MOS capacitor, shown in Fig. 6.10 is formed between the gate electrode and the conducting channel of a MOS device. The MOS capacitor is used both in inversion mode as in accumulation mode. The positive part of the voltage axis in Fig. 6.11 gives the capacitance density in inversion mode. The negative voltage axis in Fig. 6.11 gives the capacitance density in accumulation mode (Andreani and Mattisson 2000).

Characteristics The capacitance density of a MOS capacitor heavily depends on the capacitor's operation point (Samavati et al. 1998). In Fig. 6.11 the MOS capacitance is shown as a function of the capacitors bias voltage V_{cap} for different technology nodes (90–180 nm CMOS). It is clear that the capacitance density improves when

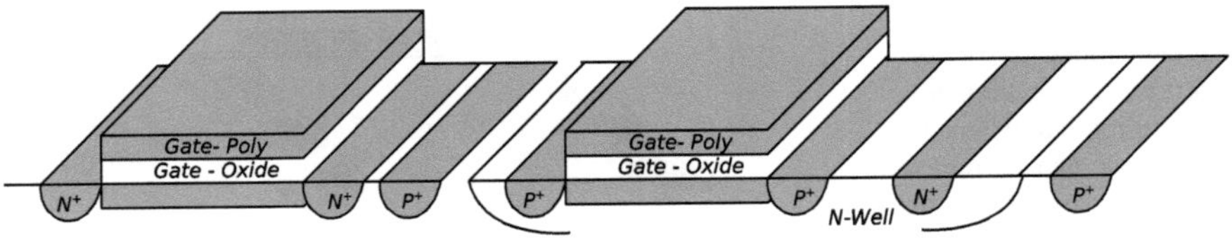

Fig. 6.10 Cross-section of the MOS capacitor's doping profiles

Fig. 6.11 Evolution of NMOS capacitance density over succeeding technology nodes and as a function of the Capacitor's operation point (Vcap)

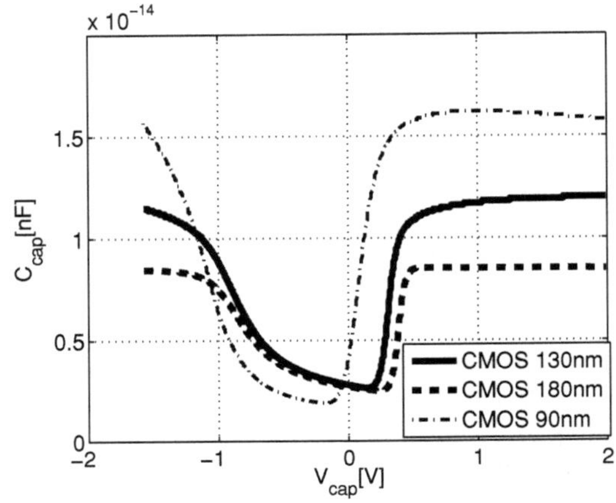

moving down the technology-nodes. But the capacitance density is relatively constant for a broad range of voltages and dramatically decreases for capacitor voltages below the technology threshold voltage (V_t). This makes MOS capacitors less suitable for use in DC–DC converters with small (sub V_t) capacitor bias voltages.

An additional effect of scaling is that the threshold voltage V_t decreases as well, this is demonstrated in Fig. 6.11 so that the capacitance density remains high even down to 0.3 V in the 90 nm CMOS node. In case the steady-state bias voltage of the capacitors is above the threshold voltage, the non-linear behavior still affects the start-up behavior of capacitive DC–DC converters. Since during start up, the capacitors need to be charged and at this phase the converter's charge rate is reduced by the limited capacitance and the start-up boundary conditions.

Moreover MOS capacitors can only be used if the channel is isolated from the die's substrate and the device's wells remain reverse biased during the converter's operation. NMOS devices obtain isolation by using triple-well structures while PMOS devices are isolated by nature. On the other hand triple-well devices suffer from larger parasitic capacitance than the PMOS capacitor. But a PMOS capacitor suffers from lower intrinsic mobility and thus the higher channel resistance.

Model In Fig. 6.12 a straightforward model of the MOS capacitor is presented. The model is characterized by a distributed resistor-capacitor ladder. The equivalent series resistance is calculated as the series resistance of the channel resistance $R_{Channel}$ and the poly-gate resistance R_{Poly}. The channel resistance is proportional to the

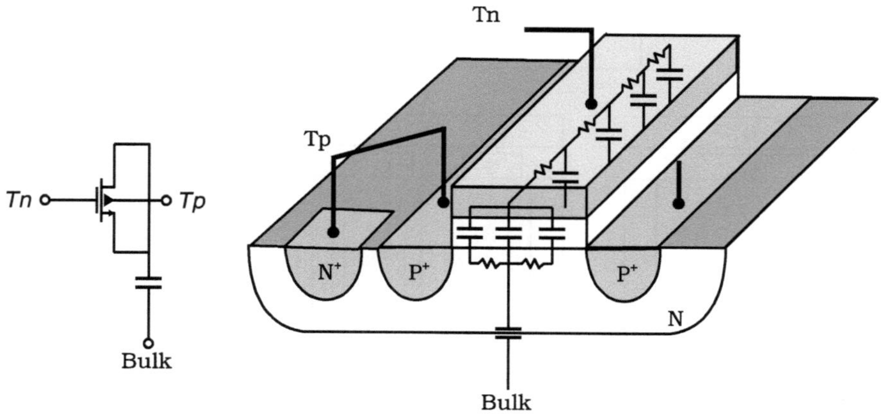

Fig. 6.12 Section of a MOS capacitor indicating the equivalent lumped model used in optimization of the equivalent series resistance of an integrated MOS capacitor

transistor's length and inversely proportional to the transistor's width:

$$R_{Channel} = \frac{L}{W} R_{Channel,\square} \tag{6.9}$$

The gate resistance is proportional to the transistor's width and inversely proportional to the transistor's length:

$$R_{Gate} = \frac{W}{L} R_{Poly,\square} \tag{6.10}$$

The eventual equivalent resistance is also reduced by connecting the channel/gate at both sides. This results in a factor $\frac{1}{12}$ (Razavi et al. 1994) which makes the eventual equivalent resistance equal to:

$$R_{ESR} = \frac{W}{12L} R_{poly,\square} + \frac{L}{12W} R_{Channel,\square} \tag{6.11}$$

To achieve a low enough R_{ESR} the switch is fragmented in small fingers and connected intensively with metal. This is demonstrated in Fig. 6.13. By fragmenting the MOSCAP, the R_{ESR} is reduced through paralleling the capacitors, but there is an obvious area penalty due to the drain/source/bulk-contact overhead. The effective amount of capacitance is decreasing while fragmenting and thus the capacitance density decreases. This demonstrates a trade-off between a low R_{ESR} and a high capacitance density. By decreasing W and L the series resistance is decreased, but at the cost of the capacitance density. This reduction in capacitance density is invoked by the increasing area overhead due to the additional source and drain area, as is suggested in Fig. 6.13. But in Fig. 6.14 it is demonstrated that fragmentation has an

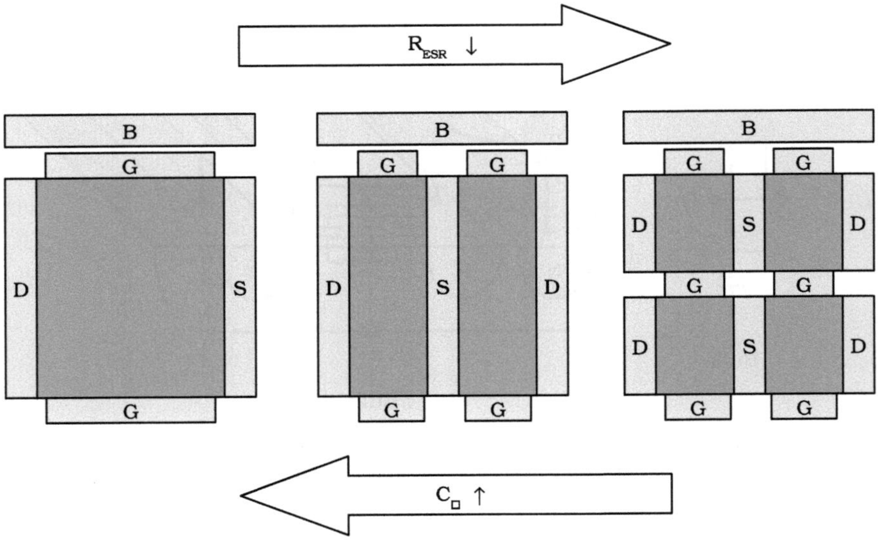

Fig. 6.13 Graphical representation of the fragmentation of an integrated MOS capacitor and the influence of fragmentation on the capacitor overhead

Fig. 6.14 Scaling of capacitors: Capacitor Quality (ratio of parasitic capacitance w.r.t. main capacitance) in function of the main capacitance for MOS capacitors, MIM capacitors between Metal 8–9 and MIM capacitors between Metal 4–5

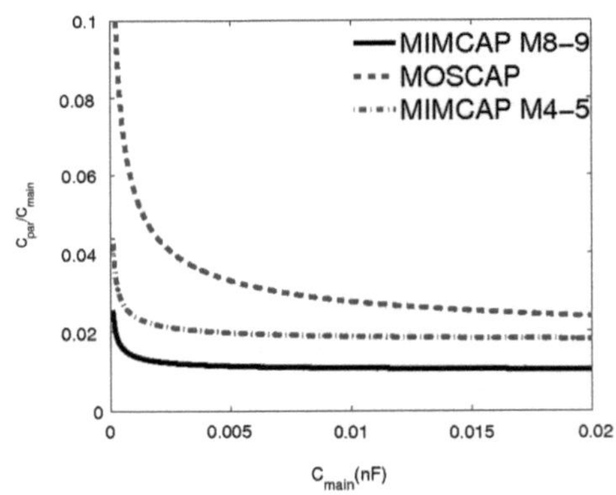

effect on the quality of the capacitors as well. But this effect is only significant when the capacitors (in 90 nm CMOS) become smaller than 10 pF.

In Figs. 6.15 and 6.16 the results of a design-space exploration of a PMOS capacitor is depicted. In both figures the capacitance density and corresponding equivalent series resistance of a capacitor with a total capacitance of 2 nF is calculated as a function of the width and length of the transistor fingers. In practice a maximum acceptable series resistance is calculated and the sizing corresponding to the highest capacitance density to achieve this resistance value is selected.

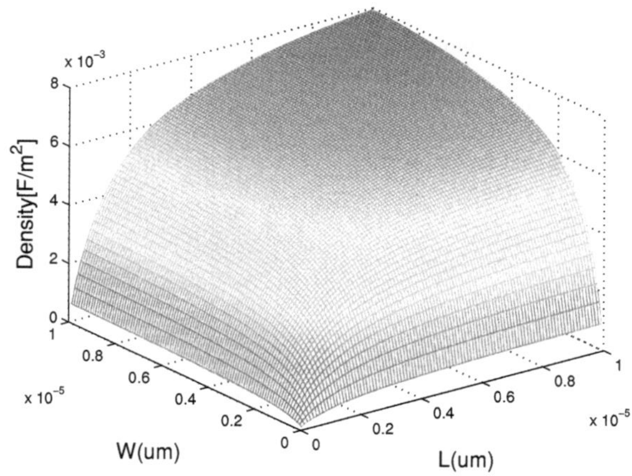

Fig. 6.15 Results from a design-space exploration for a 2 nF PMOS capacitor in a 90 nm technology: Capacitance density as a function of width and length of the capacitor's fingers

6.3.5 Technology Assessment

Switches

Under influence of the digital scaling hype, the minimum gate-oxide thickness of standard CMOS is being reduced continuously (ITRS 2009). By doing this the gate capacitance density of solid state switches is increased (Jaeseo Lee et al. 2007), at the same time the on-resistance is decreased even more, yielding a better $q_{gate}R_{ON}$ transistor figure of merit. This results in decreasing switch resistance considering a fixed transistor area. Obviously this evolution, which is driven by the digital business, is quite beneficial for designers engaged in the design of integrated power supplies (Seeman et al. 2010). They can implement switches with equally small on-resistance on smaller area and with reduced gate switching losses.

Capacitors

From Fig. 6.11 it can be seen that for MOS capacitors the capacitance density increases when migrating to technologies with smaller feature sizes. MOM capacitor's quality increases due to the increase in metal layers and the decreasing metal pitch which both improve the capacitance density and potentially reduces the parasitic capacitance if only the upper part of the metal stack is used for these capacitors. MIM capacitors slightly benefit from scaling by increasing the distance between the bottom plate of the capacitor and the substrate, because the MIM capacitors density can only be improved by using high-permittivity materials for the dielectricum and reducing the distance between the capacitor plates.

Of course not only the capacitance density is important. In Fig. 6.14 the ratio of parasitic capacitance and useful capacitance of three different types of capacitors in a 90 nm CMOS technology is given. While the capacitance density of MOS capacitors

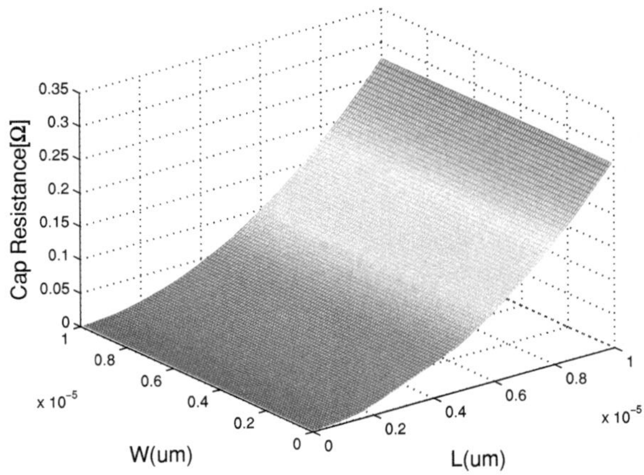

Fig. 6.16 Results from a design-space exploration for a 2 nF PMOS capacitor in a 90 nm technology: Capacitor series resistance as a function of width and length of the capacitor's fingers

is higher, the MIM capacitors demonstrate a lower capacitive coupling to the substrate. Especially the MIM capacitors in a 9-Metal stack have a very low capacitive coupling. But as already has been noticed in Fig. 6.11 scaling of CMOS technology projects a continuous improvement of the density of MOS capacitors. The latter provides the possibility to build more efficient converters with an increasing power density in DSM CMOS technologies.

6.4 Conclusion

The realization of power electronics devices in a standard CMOS technology is far from evident. Previous work (Mike Wens and Michiel Steyaert 2011) on inductive-type DC–DC converters identified the parasitic capacitance between a metal-track inductor and the equivalent series resistance of the same inductor as the main bottleneck to achieve high efficiency conversion on-chip. According to the most recent predictions these obstacles will not be removed easily (Jaeseo Lee et al. 2007) in standard CMOS. The assessment performed in this chapter demonstrates that when using capacitors instead of inductors as energy-transfer components, these issues are overcome. Especially since capacitors gain a lot from the scaling of CMOS: both capacitance density is improved and the parasitic capacitance is reduced. This provides the capacitive DC–DC converters a lot of momentum with respect to their inductive counterparts and turns them into excellent alternatives.

Chapter 7
DC–DC Converter Prototypes

This chapter gives an overview of a number of capacitive DC–DC converter prototypes. In addition to the theoretical background presented in the previous chapters, the prototypes illustrate the proposed techniques and the measurements demonstrate their performance.

7.1 Multi-Phase High-Efficiency Voltage Doubler

7.1.1 Introduction

Up-conversion DC–DC converters are especially favored in high-performance digital systems. They are used for forward-biasing body terminals,[1] for state-retention during power-gating and in memories if higher writing voltages are required Somasekhar et al. (2010). Next to a high efficiency, a small output-voltage noise (ripple) is the primary specification. Therefore a multi-phase interleaving technique[2] is adopted which is based on the fragmentation of the capacitive DC–DC converter into a large number of equivalent converters. The fragmentation facilitates an alternative switching method, that distributes the switching instants and smears out the noise in time. This chip Breussegem and Steyaert (2009) together with Somasekhar et al. (2009), is the first prototype that promotes multi-phase interleaving by means of fragmentation. The multi-phase interleaving technique has become widely adopted Le et al. (2010), Chang et al. (2010), Breussegem and Steyaert (2010), Salem and Jain (2011), Salem and Ismail (2011) and is regarded as the most effective technique to reduce ripple with a negligible cost penalty while improving power density Le et al. (2010).

[1] This technique is discussed in Chap. 2.

[2] Multi-phase interleaving is discussed in Chap. 4.

T. Van Breussegem and M. Steyaert, *CMOS Integrated Capacitive DC–DC Converters*, Analog Circuits and Signal Processing, DOI: 10.1007/978-1-4614-4280-6_7,

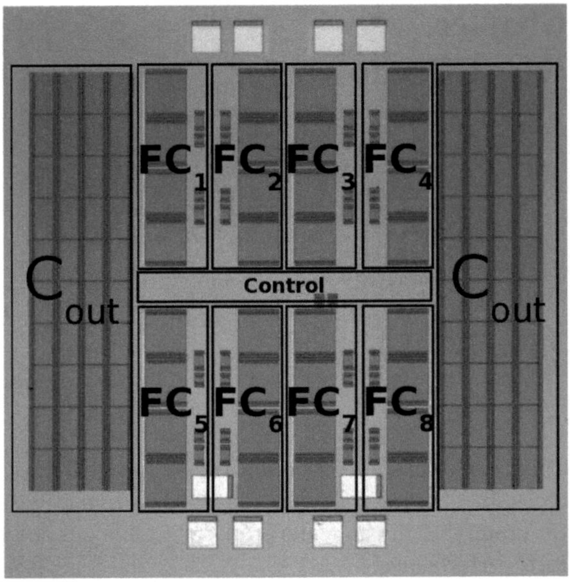

Fig. 7.1 Micro-photograph of the multi-phase voltage doubler: each block FC_i includes a separate converter core

7.1.2 Summary

A fully integrated voltage doubler is built. Thanks to a 16-phase switching scheme, the output voltage ripple is reduced to less than 0.5 % of the output voltage. The peak efficiency and the nominal efficiency are respectively 82 % and 79 %. The DC–DC converter has been fabricated in a 130 nm CMOS process and as well the charge pumping capacitors as the output buffer capacitor have been implemented by means of MIM capacitors. The maximum switching frequency reaches 20 MHz but backs-off in low-load operation in order to reduce the switching losses.

7.1.3 Converter Structure

Each converter core FC_i comprises two voltage doublers and each voltage doubler consists of four switches and two capacitors (M_1-M_2-C_x-C_y). In Fig. 7.2a two voltage doublers are shown. By crosscoupling the switches of the voltage doublers, implementation of separate level shifters is omitted. An additional charge pump (M_3-M_4-C_{xx}-C_{yy}) , added in Fig. 7.2b, is used for biasing the PMOS devices (M_2). By using this additional biasing the switch leakage is reduced. Node clk and Nclk are driven by clock drivers and these push and pull the flying-capacitors C_x-C_y up and down in an alternating way. In unloaded-conditions the capacitors are charged up to the input-voltage, so that the nodes n_x and n_y are switched between the input voltage and twice the input voltage. The PMOS (M_{2a} and M_{2b}) alternately connects

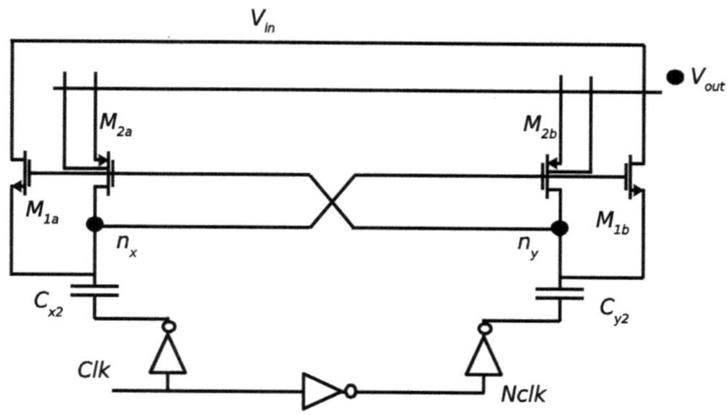

(a)

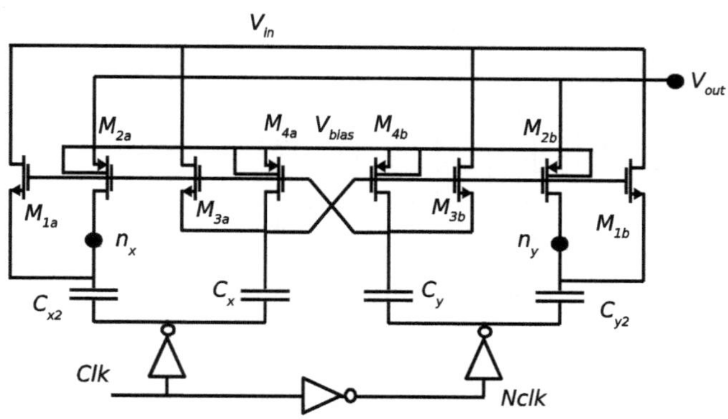

(b)

Fig. 7.2 Charge-Pump structure **a** A simple charge-pump structure with cross-coupled switch drive **b** A charge pump with additional charge pump to reduce the switch reverse leakage current

the output-capacitor to the nodes n_x and n_y. These nodes are both charged to twice the input-voltage so that the output-buffer capacitor is continuously charged to twice the input voltage. Under loaded-conditions the output voltage is lower, due to the converter's finite output impedance, however the operation principle is identical.

7.1.4 System

Chapter 4 discusses how the noise or ripple of the power supply jeopardizes the performance of electronic applications. Hence the ripple at the output of a DC–DC converter, which is generating a secondary power supply, must be limited to acceptable levels. According to Sect. 4.4, this ripple can be reduced by increasing

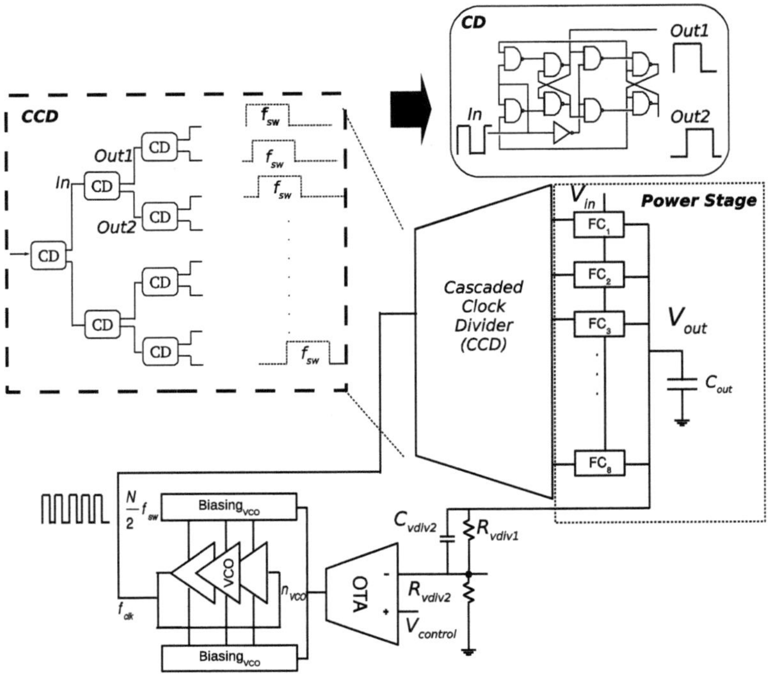

Fig. 7.3 System architecture of the 16-phase interleaved voltage doubler

the output buffer capacitance or by using a series linear regulator. Since in fully integrated converters the capacitor-size is limited, this renders big output-capacitors impossible to use. The use of a series linear regulator is area consuming and is not energy-efficient if ripple at the input node of the LDO is high. By paralleling multiple DC–DC converters and activating each converter phase-shifted with respect to the other converters, as discussed in Sect. 4.4, the output ripple is decreased. If N converters are paralleled, the switching signal of every converter will be shifted by $\frac{2\pi}{N}$.

A capacitive DC–DC converter is mainly used under closed-loop regulation. At steady-state, the output voltage remains constant under varying loads. To achieve this, the switching frequency is dynamically controlled by a control loop as is demonstrated in Chap. 5. For a constant input voltage, the ratio between actual and ideal output voltage of the converter γ, is kept constant by means of closed-loop frequency-regulation.

The closed-loop regulation, based on the lead-compensation technique from Sect. 5.3, is established as is shown in Fig. 7.3. The output voltage of the DC–DC converter is measured through the resistive-divider (Rvdiv1 and Rvdiv2) in order to reduce the required input swing of the difference amplifier. This difference-amplifier is a symmetrical OTA with PMOS input difference pair and rail-to-rail output. This

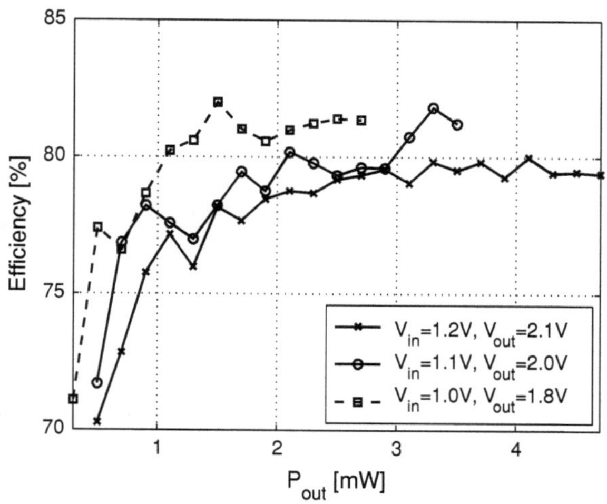

Fig. 7.4 Efficiency of the voltage doubler as a function of the output power for three distinct conversion pairs

OTA regulates the switching frequency which is generated by a current-starved oscillator.

The Cascaded Clock Divider (CCD), shown in Fig. 7.3, generates 16 overlapping clock phases to provide the multi-phase clocking scheme. This circuit is built from a Clock Divider (CD) circuit that generates two clock signals at half the frequency of the input signal, but shifted over 0 and 90 degrees. By cascading this CD in 3 levels, 2^3 signals are generated. The loop stability is guaranteed by a loop-compensation capacitor of 666 fF.

7.1.5 Measurement Results

The performance of the voltage doubler has been measured under a range of different loads and input voltages.(V_{in}: 0.65–1.2 V, V_{out}: 1.2–2.1 V, R_{load}: 10 k–2 k) and the performance is similar for all conversion pairs: efficiencies between 70 % and 82 % (Fig. 7.4). Ripple smaller than 10 mV (Fig. 7.5). The performance for a 1.1–2.0 V conversion (200–808 μA load variation) efficiency between 71.3 % and 79 %, a ripple between 4.5 mV and 6.4 mV.

]

For qualifying the ripple reduction irrespective of the DC–DC converter characteristics a FOM is defined. In Sect. 4.2 the ripple of a simple two-phase charge-pump is defined as in:

$$\Delta v_{SSL,simple} = \frac{I_{load}}{C_{out} \times f_s} \tag{7.1}$$

The ripple of such a capacitive DC–DC converter, operating in the SSL, is proportional to the load current and inverse proportional to the output capacitor size

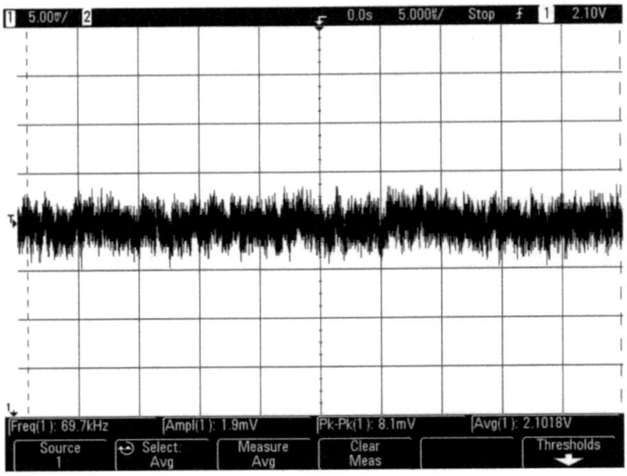

Fig. 7.5 Ripple detail at an input voltage of 1.2 V and output voltage 2.1 V at nominal load

Table 7.1 Comparison of the designed converter and known converters in literature

Work	Iload [A]	Cout [F]	Ripple [V]	fs [MHz]	FOM
1: Cheng et al. (2008)	3.00E-002	1.00E-005	1.00E-002	1.50E+005	0.50
2: Lee et al. (2006)	3.00E-002	2.00E-006	3.30E-002	6.00E+005	1.32
3: Wu and Chen (2008)	2.00E-002	1.00E-006	2.80E-002	1.00E+005	0.14
4: Lau et al. (2007)	1.50E-001	2.20E-006	3.00E-002	2.00E+005	0.09
5: TVB	1.00E-003	4.00E-010	1.00E-002	2.00E+007	0.08

and the switching frequency. A fair FOM compares the output voltage ripple of an adapted charge-pump with the ripple produced by a simple two-phase charge pump with the same switching frequency, output capacitor and delivering the same current to the load:

$$FOM = \frac{\Delta v_{SSL}}{\Delta v_{SSL,simple}} \tag{7.2}$$

This FOM expresses the fraction to which the ripple was reduced. Table 7.1 gives an overview of the state-of-the-art voltage doublers found in literature. It is clear that Cheng et al. (2008) and Lee et al. (2006) do not demonstrate significant improvement with respect to a straightforward implementation. Although these converters use a large output capacitor, they don't succeed to reduce the ripple substantially. The FOM of Lee et al. (2006) is bigger than one because some non-idealities (f.e. Resr of Cout and connection resistance) cause the ripple to be higher than the ripple defined in Lee et al. (2006) which does not take these non-idealities into account. The multi-phase capacitive DC–DC converter presented in this section outperforms all known fully integrated voltage doublers published before 2009.

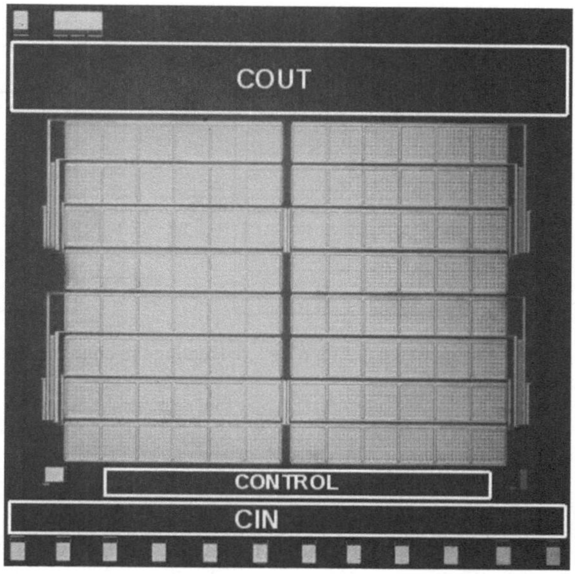

Fig. 7.6 Micro-photograph of the reconfigurable low-power hysteretic DC–DC converter prototype. The chip measures 1.8 × 1.8 mm, including all bond pads and a seal ring

7.1.6 Conclusion

This fully integrated multi-phase capacitive DC–DC converter operates at efficiencies between 70 % and 82 %. The charge pump itself is proven to work equally efficient over a broad range of input and output voltages (Vin:0.65–1.2 V, Vout:1.2–2.1 V). It is built for small-ripple purposes. The interleaving strategy reduces the output-voltage ripple down to 2 % of the ripple of an straightforward two-phase charge pump.

7.2 Reconfigurable Hysteretic DC–DC Converter

7.2.1 Introduction

The main issues faced by low-power monolithic capacitive DC–DC converters are the topology-dependent iVCR and the limited power budget for implementing the control loop. The relationship between the topology and the iVCR is alleviated by implementing multiple topologies and thus on-line reconfigurability as discussed in Sect. 2.3. To overcome the limited power budget, alternative control techniques are required. Hence the fully digital hysteretic controller is presented in Chap. 5. This prototype demonstrates the implementation of both the multi-topology approach and the fully digital hysteretic controller (Fig. 7.6).

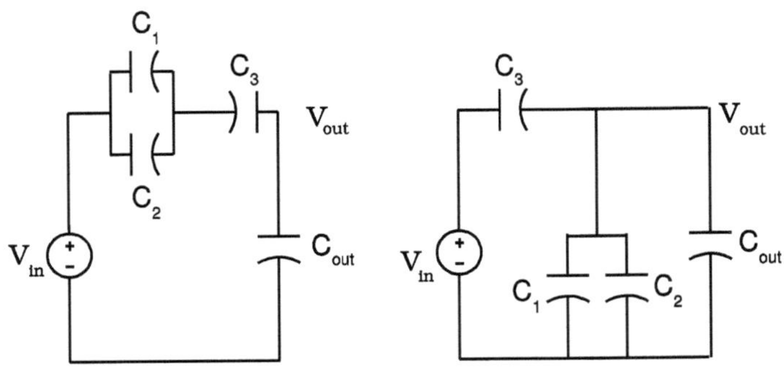

Fig. 7.7 A fractional 2/3 topology, presenting the configurations during both states of the conversion: Topology I

7.2.2 Summary

A fully integrated capacitive DC–DC converter with a multi-topology structure has been built. By merging two topologies (a fractional 4/5- and 2/3-topology) into a single structure, the output-voltage range is increased. Moreover a dual-loop digital controller improves load regulation, compared to a conventional hysteretic control technique. The converter has been implemented in a 90 nm CMOS technology with integrated MIM capacitors and achieves a peak efficiency of 88 % with an output power range from 20 up to 4.9 mW.

7.2.3 Converter Structure

Chapter 2 establishes a broad range of possible topologies to achieve an iVCR. Given a required iVCR these iVCRs can be implemented both by series-parallel topologies and fractional topologies. In Fig. 7.7 a fractional 2/3-topology is presented, in Fig. 7.8 a 2/3 series-parallel, in Fig. 7.10 a 4/5 fractional topology and in Fig. 7.9 a 4/5 series-parallel alternative. In each figure both states of these two-state converters are represented. The series-parallel topologies are quite commonly used topologies Kwong et al. (2009), Seeman et al. (2008) while topologies I and IV are newly proposed as alternatives in Breussegem and Steyaert (2010).

When implementing two different topologies in a single converter some considerations are of primary concern. First efficiency of the topologies, secondly what is the overhead that is invoked by combining two topologies. The efficiency of the topologies can be compared based on the switch- and capacitor- charge vectors and their aggregate quadratic sum: K_S and K_C. The switch overhead depicts the number of additional switches necessary when adding an additional topology. The switch overhead is expressed as a percentage of the switches, which are not in common for both topologies, with respect to the total number of switches.

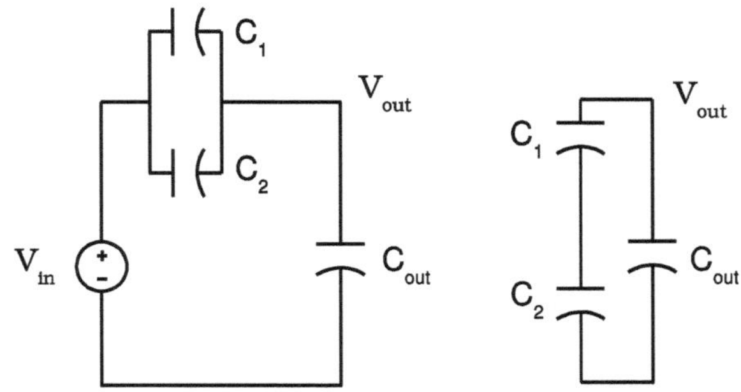

Fig. 7.8 A series-parallel 2/3 topology, presenting the configurations during both states of the conversion: Topology II

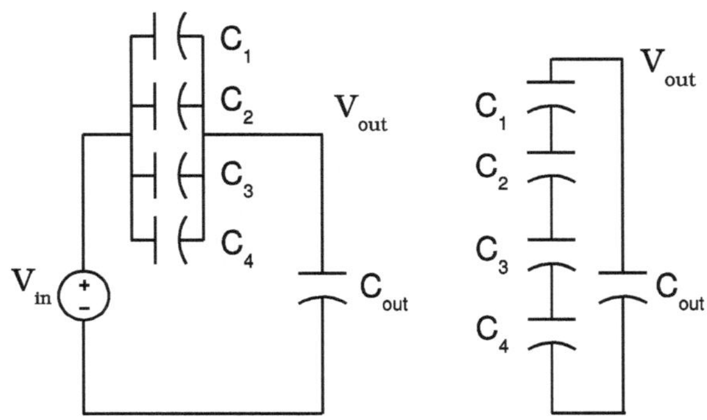

Fig. 7.9 A series-parallel 4/5 topology, presenting the configurations during both states of the conversion: Topology III

One can see that Topology I and II have equal output impedance for equal switch size and switching frequency constraints and also Topology III and IV perform equally. Topology I and II have $K_C = 0.44$ and $K_s = 5.44$, topology III and IV have $K_C = 0.64$ and $K_s = 6.76$. This means that from a performance point of view any combination of Topology I-IV or I-III or II-IV or II-III is equivalent.

The next analysis is made concerning the correspondence in topology between the candidate topologies. This will reflect in the number of extra switches that are not common for both topologies. This will not necessarily affect the efficiency or power output but will have a negative effect on the area necessary for implementing the gear-box and the control complexity. The following observations are made: Combining Topology I and IV leads to 3 extra switches on a total of 12 switches. Combining Topology II III leads to 5 extra switches on a total of 16 switches. Combining Topology II and IV leads to 10 extra switches on a total of 18 switches. Combining Topology I and III leads to 13 extra switches on a total of 20 switches. It is clear that the multi-topology converter, shown in Fig. 7.11, built from Topology I and IV has a considerable area benefit compared to a combination of the other topologies. This will not only be reflected in an reduced area but as well in less complex clocking schemes. Especially the presence of a common phase will lead to simplified control implementation.

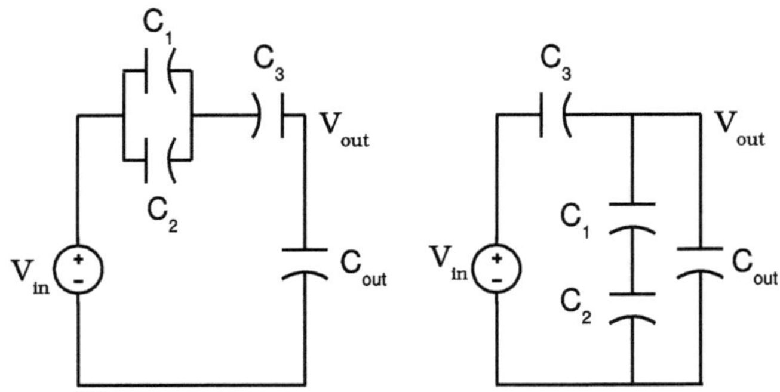

Fig. 7.10 A fractional 4/5 topology, presenting the configurations during both states of the conversion: Topology IV

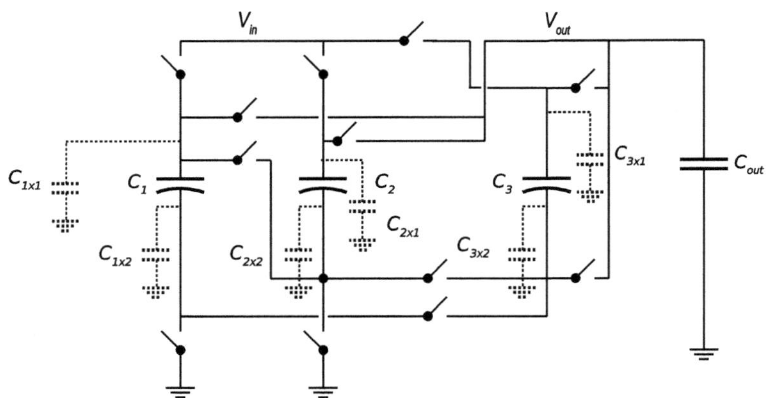

Fig. 7.11 Implemented converter structure, including parasitic capacitance due to the switches

7.2.4 System

This prototype revealed that for low-power DC–DC converters the control loop consumes a major portion of the power budget. This results in significant reduction of the power efficiency but also a steep efficiency roll-off in case of a low output power. Especially the use of analog building blocks forms the bottleneck. On the one hand a large gain and bandwidth is required to achieve low static offset in combination with fast transient behavior. On the other hand a large gain-bandwidth requires large bias currents and introduces a large power loss. Digital techniques offer considerable benefit with respect to the conventional analog techniques. First bias currents are eliminated and moreover the use of comparators instead of operational amplifiers can increase the bandwidth. Therefore a dual control method is implemented shown in Fig. 7.12. It consists of a frequency-modulation loop by means of a digital and

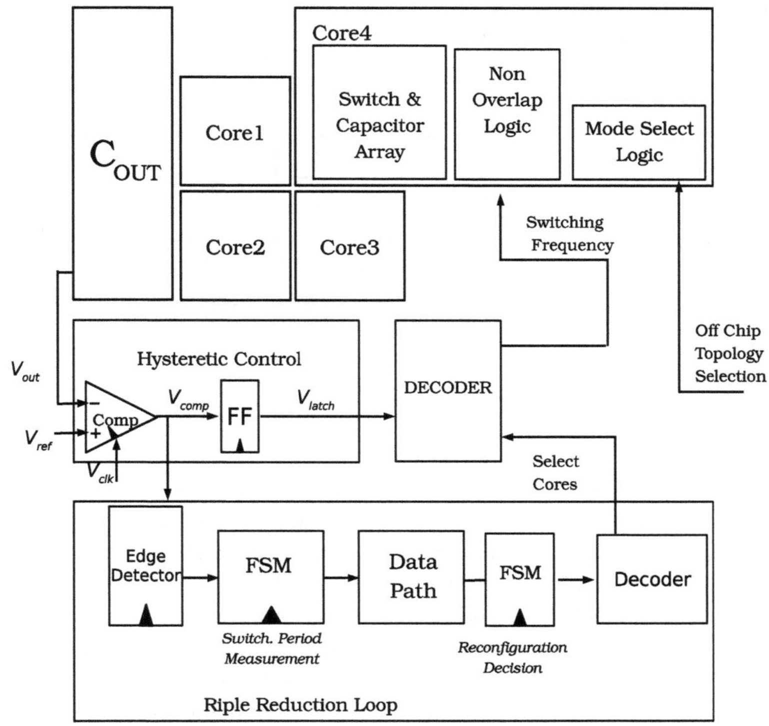

Fig. 7.12 Schematic overview of the reconfigurable multi-topology DC–DC converter with two control loops: hysteretic control and a digital Ripple-Reduction Loop

hysteretic controller and a digital loop that intervenes to reduce the ripple. Both loops are discussed in the next paragraphs.

Hysteretic Control

The hysteretic controller, shown in Fig. 7.12, is implemented as is discussed in Chap. 5. It comprises a clocked comparator and a flip-flop. The comparator observes the output voltage and by comparing the converter output voltage to the control/reference voltage, the clock signal is either passed or not. When the output voltage at V_{out} is lower than the control voltage at V_{ref}, as is demonstrated in the timing diagram in Fig. 7.13, the signal V_{comp} is passed to the succeeding flip-flop. The flip-flop filters out the falling edges of the clock signal and halves the frequency of the signal V_{latch}. This block must be incorporated since only the rising edges of the signal coming from the comparator correspond to a so-called boundary violation, which is discussed in Sect. 5.3.

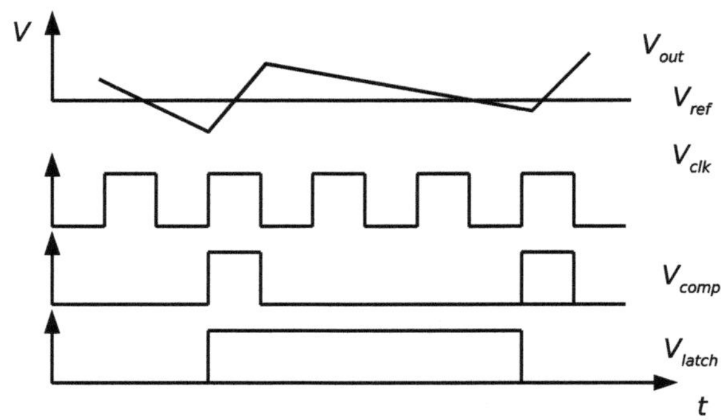

Fig. 7.13 Timing diagram of the hysteretic controller in the reconfigurable converter

Ripple-Reduction Loop

Chapter 4 demonstrates that at low load current the ripple is higher than at high load current, due to the low switching frequency and the decreased influence of the parasitic channel resistance of the switches. Therefore it would be beneficial that the switching frequency would be high at low loads as well. Higher switching frequencies imply however that the delivered power is also higher thus a technique to increase the switching frequency and keeping the delivered charge constant is required. This is accomplished by reducing the amount of flying capacitance according to the required increase in switch frequency. By fragmenting the charge pump in four equivalent paralleled charge pumps, parts of the charge pump can be de-activated for compensating the switching frequency increase.

For achieving capacitance modulation a digital loop, shown in Fig. 7.12, is added. called the Ripple Reduction Loop (RRL). It functions according to a capacitance modulation technique, which is discussed in Sect. 4.4, but in this case it is not used to perform load regulation Ramadass et al. (2010) but to modify the converter's structure so that the hysteretic loop performance is improved.

The RRL, demonstrated in Fig. 7.12, is implemented without interfering with the hysteretic control loop. The switching frequency is measured by means of two Finite State Machines (FSM). These FSM's detect a low frequency and decide (hard coded) whether a higher frequency is feasible. The FSM's will deactivate part of the charge pump and therefore the hysteretic controller will increase the switching frequency.

If the switching frequency rises and crosses a hard-coded boundary the FSM's will detect this and activate the different charge pumps one by one until the switching frequency crosses the frequency boundary again and enters the safe zone.

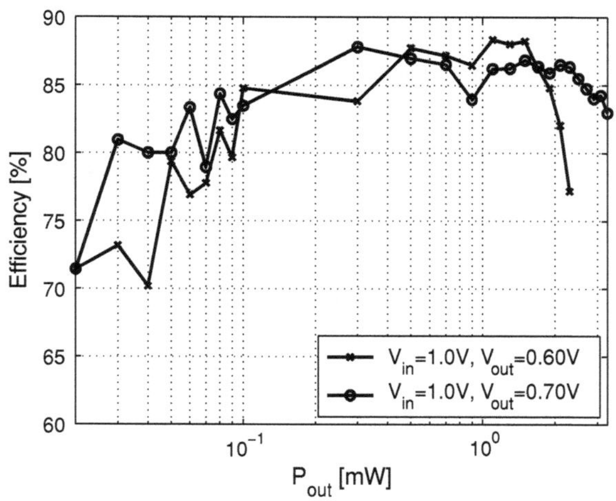

Fig. 7.14 Efficiency measurements of the multi-topology converter at Vin = 1.0 V

7.2.5 Measurement Results

Efficiency

It has been explained in previous sections that the converter (shown in Fig. 7.1) has two different modes corresponding to distinct topologies. Each mode has been characterized separately. Efficiency has been measured for both nominal input voltages: 1.2 V and 1.0 V. The output voltages corresponding to the VCR's that are implemented in the multi-topology converter are 0.85 V/0.72 V for the 1.2 V input and 0.7/0.6 V for the 1.0 V input. The measurements were performed with a constant external current load.

In Fig. 7.14 the measurements are shown for the 1 V input case and the 1.2 V input case is shown in Fig. 7.15. Efficiency remains above 80 % for a 300–4.9 mW load. But above respectively 80–40 μW efficiency is better than an ideal linear regulator for resp the 4/5 and the 2/3 mode. When a 1.0 V is applied one can see that the DC–DC-converter performs better than an LDO over the power range down to 30 μW. A peak efficiency of 87 % is achieved 2.7 mW from a 1.2 V input.

Varying Output Voltage

Because of the multi-topology principle, high efficiency is maintained over an extended output voltage range this is demonstrated in Fig. 7.16. The converter is measured for a 1.2 V input voltage and a 3mA load. The control voltage is swept so that the output voltage varies between 0.85 V and 0.5 V. With a single topology the efficiency remains above 80 % between 0.85 V and 0.75 V, thanks to the merged topology-gear box approach this range is extended down to 0.65 V. Over the whole

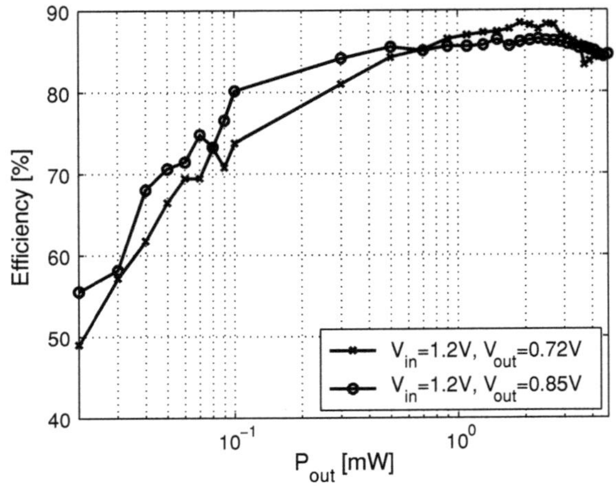

Fig. 7.15 Efficiency measurements of the multi-topology converter at Vin = 1.2 V

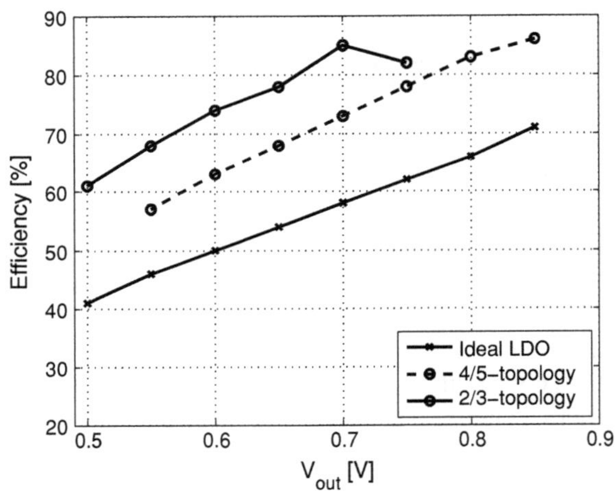

Fig. 7.16 Efficiency of the converter subject to constant load and varying control voltage

range, the merged topology approach remains at least 10 % above the maximum efficiency of an ideal linear regulator. In Fig. 7.17 the Efficiency Enhancement Factor (EEF) is plotted for this case. A maximum EEF of 33 % is achieved.

Varying Input Voltage

This converter functions properly for voltages down to 0.7 V. This is demonstrated in Fig. 7.18. For an input voltage range between 0.7 V and 1.2 V, a load current of 1 mA and in the 4/5-mode, efficiency remains higher than 80 %.

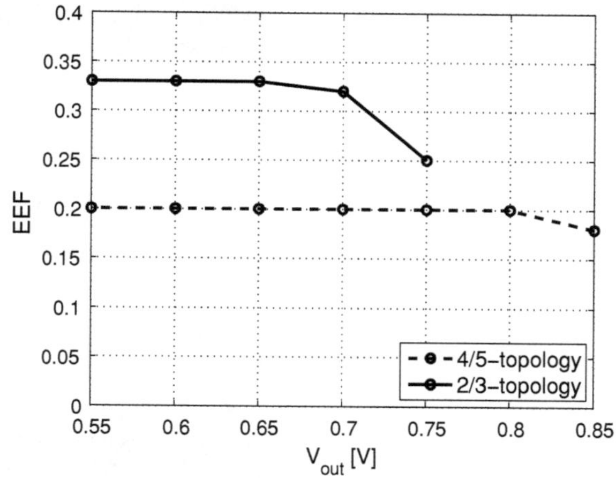

Fig. 7.17 Efficiency Enhancement Factor of the converter subject to constant load and varying control voltage

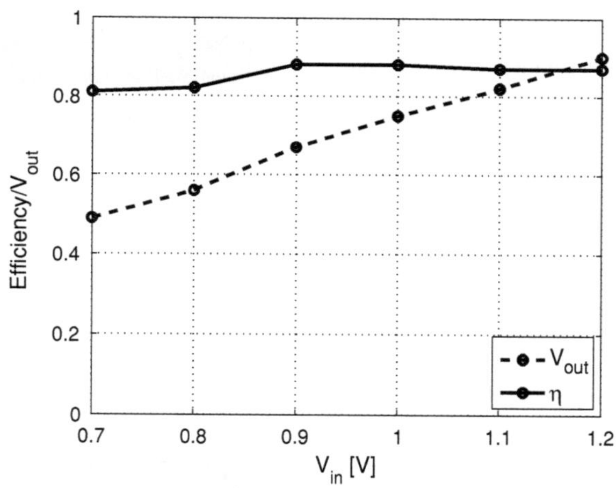

Fig. 7.18 Efficiency and output voltage as a function of varying input voltage at constant γ

Load Regulation

For both the 4/5- and the 2/3- mode, a load step was applied at 1 MHz. The load step varied the load current between 1.86 mA and 0.186 mA. This is done once while the RRL was bypassed and once when the RRL was activated. The measurement of the 4/5 Mode is shown in Fig. 7.19 and Fig. 7.20.

In both cases very fast regulation is achieved. In case the RLL is bypassed, the response to the load change was quasi instantaneous taking less than 10 ns. The influence of the load step on the RMS output voltage is not negligible: 20 mV. By activating the RRL, the recovery time increases to up to 50 ns but the load regulation is much better. The change in RMS voltage was less than 5 mV. This demonstrates the benefits of the RRL not only for the reduction of ripple but also for the load

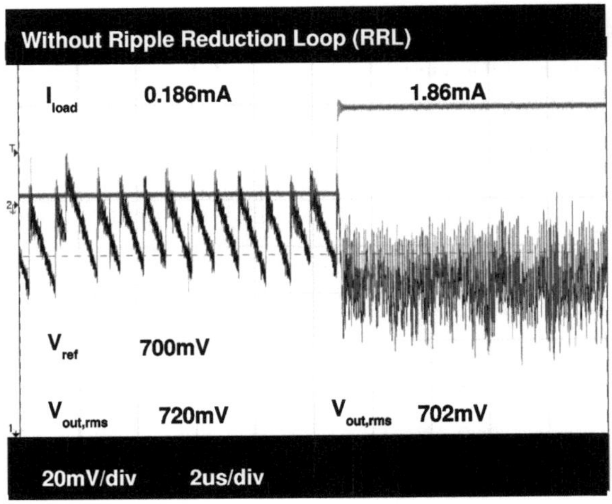

Fig. 7.19 Effect of a load step on the converter output voltage without RRL

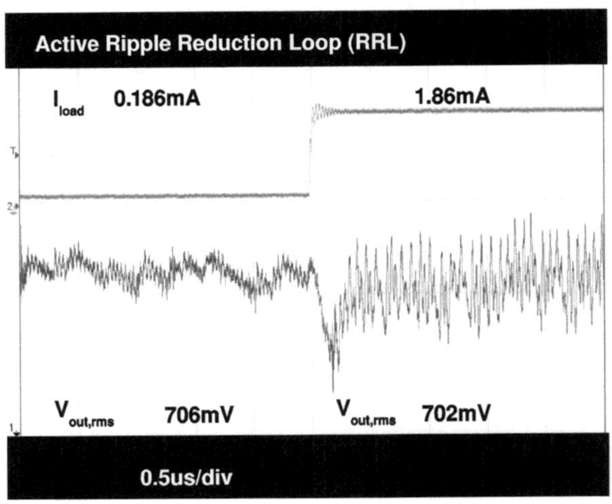

Fig. 7.20 Effect of a load step on the converter output voltage with RRL active

regulation specifications. Power measurements prove that the use of the RRL has no influence on the power conversion efficiency of the system.

The effectivity of the RRL can be observed in the first place by analyzing the switching frequency under RRL regime and regular regime. If we analyze the 2/3 mode, one can see that the RRL reduces the ripple by a factor of 2 by means of increasing the switching frequency and thus deactivating half of the active charge pump compared to the bypassed regime. Power conversion efficiency is not affected by activating the RRL.

7.2.6 Conclusion

A fully integrated capacitive DC–DC-converter with multiple topologies is presented. By merging these topologies into a single switch-capacitor array the output voltage range is increased. The dual-loop digital control improves load regulation compared to a conventional hysteretic control and reduces ripple under low-load operation.

7.3 Single-Boundary Multi-Phase Hysteretic Converter

7.3.1 Introduction

Monolithic integration in CMOS has boosted the development of low-cost, compact and portable consumer applications. But for a long time the monolithic integration of DC–DC converters is still omitted in commercial applications, primarily due to the need for high-efficiency converters and appropriate techniques to control high-frequency DC–DC converters. Therefore it is an obvious step to investigate low-power control techniques especially for medium power (>100 mW) multi-phase interleaved capacitive DC–DC converters. A new technique is developed to control multi-phase capacitive DC–DC converters in an energy-efficient way. By combining the multi-phase interleaving with hysteretic control, a high bandwidth is combined with low noise and low power losses.

7.3.2 Summary

A fully integrated capacitive step-down DC–DC converter in 90 nm CMOS with an output power capability of 150mW is presented. A peak efficiency of 77 % and a full load efficiency of 74 % has been measured. The DC–DC converter is controlled by a multi-loop multi-phase control loop. This discrete-time control method provides a low-power solution for controlling multi-phase capacitive DC–DC converters without compromising the control-loop bandwidth (Fig. 7.21).

7.3.3 Converter Structure

The single-flying-capacitor voltage divider is shown in Fig. 7.22a. It is a capacitive 1/2 series-parallel step-down DC–DC converter. The converter operates optimally for a VCR near 1/2 and this optimum is imposed by the topology of the DC–DC converter. This topology consists of a charge-transferring capacitor (the flying capacitor C_{fly}), a buffer capacitor C_{out} and four switches. The DC–DC converter

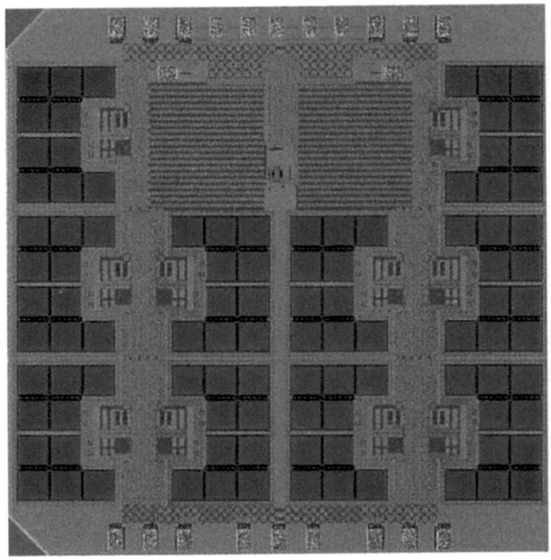

Fig. 7.21 Micro-photograph of the chip: 1.8 mm × 1.8 mm including bond pads and the ESD protection

operates in a two-state cycle. In the first state ϕ_1, shown in Fig. 7.22b, the flying capacitor C_{fly} is connected between the converter's input and its output node. During the second state ϕ_2, C_{fly} is connected between the output and the ground node (Fig. 7.22c). By switching between these configurations charge is transferred from the input to the output of the DC–DC converter.

7.3.4 System

This section examines the implementation of the fully integrated DC–DC converter. First the structure of the converter core is discussed and secondly the control loop building blocks are presented.

Converter Core

In Fig. 7.24 the transistor implementation of the converter core is depicted. The structure consists of four switches $M_1 - M_4$ and a flying capacitor C_{fly}. The output buffer capacitor is omitted in this figure. Switch M_2 and M_4 require a gate swing between the ground and the output voltage of the system, they reside in the lower voltage domain of the converter (Voltage Domain 2). M_1 and M_3, on the other hand, need a gate swing between the input voltage and the output voltage and they reside in the upper voltage domain of the converter (Voltage Domain 1). By stacking the voltage domains on top of each other and ensuring that in both voltage domains the voltage restrictions are guaranteed the input voltage ($3V < V_{in} < 3.9$ V) of the circuit can exceed the maximum voltage of the technology (2.5 V). Level shifters

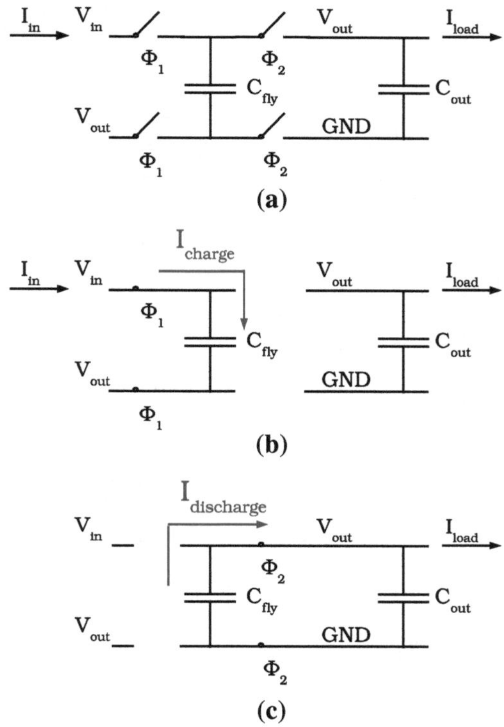

Fig. 7.22 DC–DC converter topology **a** Schematic representation of the structure **b** Charge flow during phase Φ_1 **c** Charge flow during phase Φ_2

are used to translate the control signals from the lower voltage domain to the upper voltage domain. M_3 is implemented by means of a triple-well device to isolate the bulk from the substrate. The substrate is biased with the generated output voltage. Because of this the threshold of the device was reduced and the conductance increased compared to the grounded bulk solution. For generation of the level-shifted control signals, a capacitive level shifter has been implemented.

Control Loop

Multi-loop multi-phase control is basically achieved by implementing a separate control loop for every converter core and synchronizing every control loop with N_{MP} out-of-phase clock signals. The phase-shifted clock signals are derived from a 5-tap current-starved oscillator (shown in Fig. 7.23). This oscillator generates 5 clock signals and their inverted signals. These 10 signals are used as timing signals for a tenfold comparator array. The comparator samples the output voltage through a resistive divider and evaluates the actual output voltage. The switching frequency of the converter is made independent of the comparator clock frequency by putting an edge-triggered latch between the comparator and the driver chain. The maximum switching frequency is thus half the sampling frequency. Thanks to this latch the

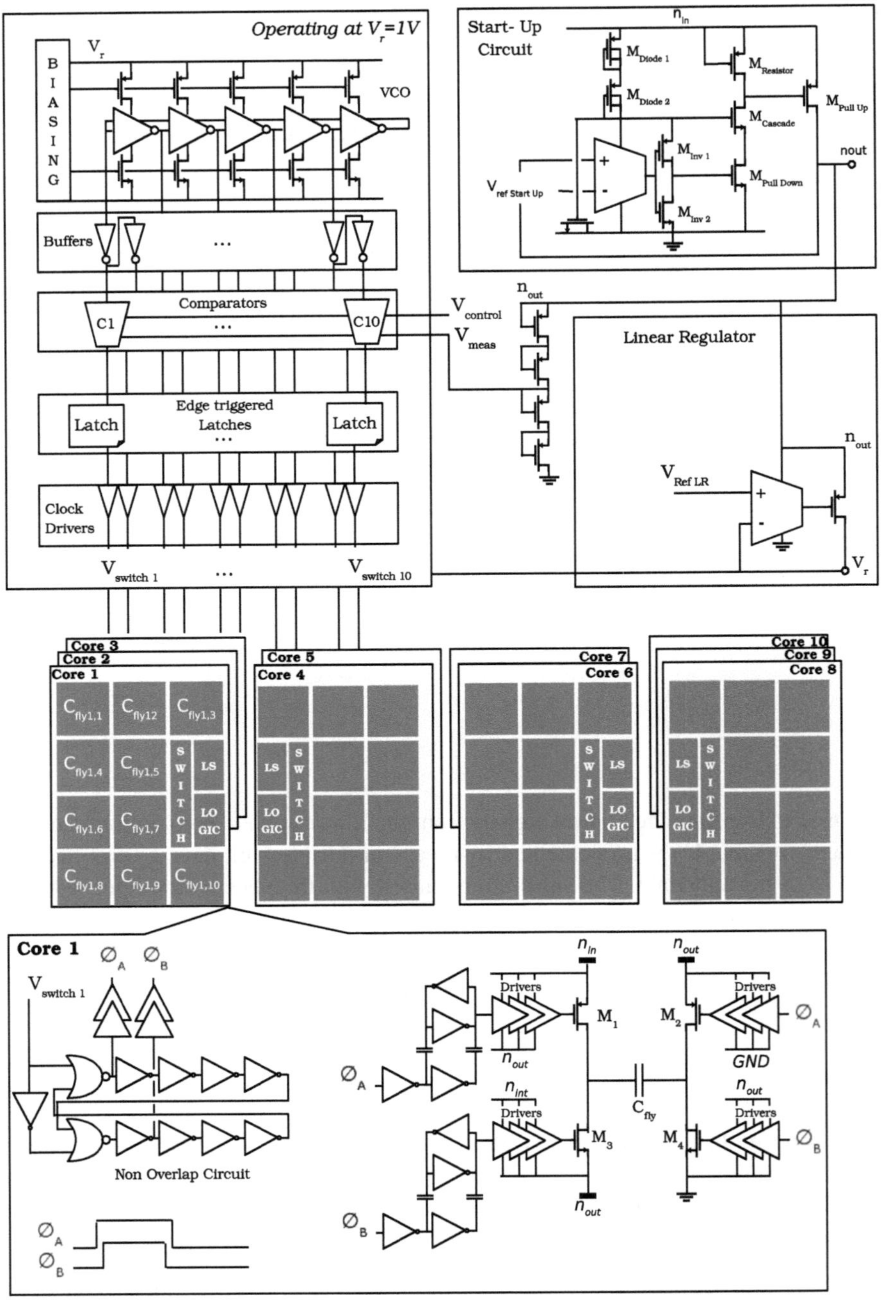

Fig. 7.23 Implementation of the multi-loop multi-phase control. The control-loop supply is derived from the DC–DC converter's output and the start-up circuit initiates the voltage ramp up. The DC–DC converter consists of 10 equivalent converter cores. Each of them is controlled by the digital-hysteretic controller

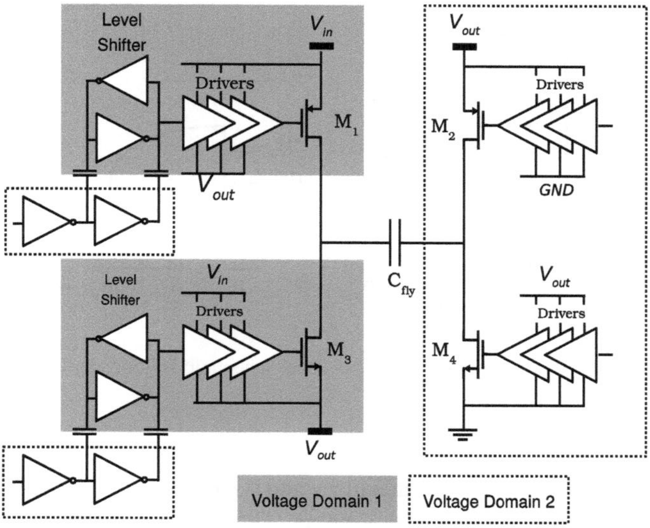

Fig. 7.24 Circuit implementation of the converter core. The upper voltage domain is marked dark and the lower voltage domain is marked by the dashed line. Only the MIM-capacitors are shared

duty cycle of the switching signal remains 50 % in steady state and charge transfer is optimal. This straightforward control loop can be used because of the dual-phase charge transfer nature of this type of converter. At each rising edge of the sampling clock an evaluation is made and if necessary the converter core toggles the phase, which implies a charge transfer to the load. Since the signals are time-shifted with respect to each other, the output voltage is sampled at a rate 10 times higher than the maximum switching frequency. And each positive evaluation is followed by a charge transfer of a single converter core. This leads to a reduced current impulse and significant noise reduction at the converters' input and output nodes. This approach reconciles the multi-phase technique with the hysteretic control technique, which is used in the previous prototype, and provides a multi-phase control of the output voltage.

For sake of efficiency a low-power minimum-sized clocked comparator (shown in Fig. 7.25) is used. The comparator operates as a triple-input analog NAND-gate. It has a digital input clk_x and two analog inputs n_{inn} and n_{inp}. And one digital output n_{comp}. The following logic operation is executed: If $n_{inp} > n_{inn}$ than $n_{comp} = \overline{clk_x}$ else $n_{comp} = 1$. So the clock is passed through based on the analog inputs. The comparison is only executed when clk_x has a rising edge. For sensing the output voltage and shifting it within a safe range (thus smaller than the breakdown voltage of the comparator's devices) we use a sub-threshold MOS resistance ladder. This requires the use of low leakage devices in the comparator input pair since the gate leakage of SP devices would corrupt the resistive voltage division of the ladder. At room temperature a resistance of 35 MΩ is achieved with a single transistor.

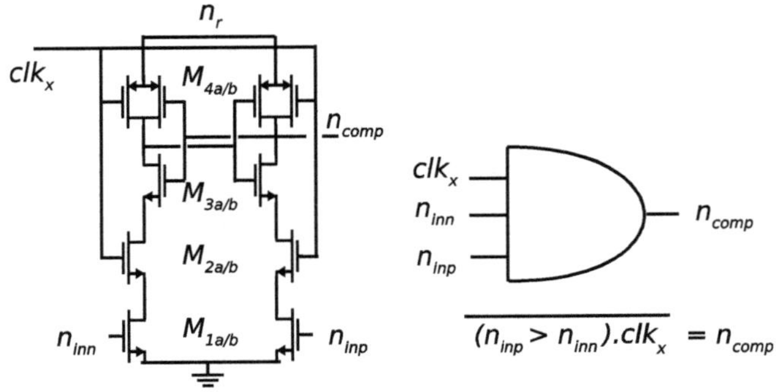

Fig. 7.25 The clocked comparator **a** Transistor level schematic **b** Equivalent Circuit of the comparator

The control-loop circuitry operates at 1 V and is fed by a dedicated LDO (Fig. 7.23) connected to the DC–DC converter's output. Thus an LDO converts the generated output voltage down to this level and a self-contained supply is generated. Simulations demonstrate that the control-loop power consumption, including level shifters and comparators, is smaller than 250 μW for full-load operation and smaller than 100 μW for no-load operation. These simulations are confirmed by the measurements in Sect. 7.3.5.

7.3.5 Measurement Results

In Table 7.5 the main specifications of the DC–DC converter are summarized and the next paragraphs will demonstrate the performance of this converter in terms of efficiency, efficiency enhancement, load regulation, variable input efficiency and start-up behavior.

Efficiency

The DC–DC converter's chip photograph is shown in Fig. 7.21. Efficiency was measured for input voltages of 3.0 V, 3.3 V and 3.6 V and output voltage of respectively 1.305 V, 1.42 V and 1.5 V. This is shown in Fig. 7.26. A peak efficiency of 77.3 % is achieved for a voltage conversion from 3 to 1.3 V resulting in an output power of 70 mW. The maximum output power for all conversion pairs was 150 mW at a maximum switching frequency of 70 MHz. From 10 to 150 mW the efficiency remains higher than 70 %. Therefore, this design proves that in Bulk CMOS as well high efficiencies can be achieved and power-dense DC–DC converters can be designed. A no-load power consumption of 85 μA has been measured for $V_{in} = 3.3$ V. Stability

Table 7.2 Performance overview

Process	90 nm UMC 1 Poly 9 Metal CMOS
$iVCR$	1/2
# Phases	10 times Multi-phase
V_{in}	3.9–3.0 V
V_{out} Regulated	1.3 at Vin 3.0
	1.42 at Vin 3.3
	1.5 at Vin 3.6
C_{fly}	2 nF MIM-cap
C_{out}	3.2 nF MOS-cap
$f_{sw,max}$	70 MHz
η_{max}	77 %
$P_{out,max}$	150 mW
$\eta_{P_{out,max}}$	74 %
EEF_{max}	44 %
$I_{in,noload}$	85 μA
Load Reg.	0.6 %/mA

is demonstrated for an output power range from 150 mW down to 0 μW, taken into account that at no load the control loop is drawing current from the output (Table 7.2).

Efficiency Enhancement Factor

In Sect. 1.3 a benchmark for switched-mode down converters is introduced: the Efficiency Enhancement Factor. This benchmark compares the efficiency of a converter η_{Conv} to the maximum efficiency an ideal linear regulator can achieve, η_{Lin}. The EEF is formulated in Equation (7.3) and calculated for the presented design in Fig. 7.27. It is clear that a converter with a negative EEF can not compete with a linear regulator. The EEF remains between +39 % and +45 % over the whole power range from 10 mW and 150 mW. Compared to an ideal linear regulator, this design performs at least 39 % better.

$$EEF = 1 - \frac{\eta_{Lin}}{\eta_{Conv}} \tag{7.3}$$

Load Regulation

The load regulation resides between −1Ω and −1.5Ω. Hysteretic control anticipates on a load variation of 30mA with a rise time of 25 ns. This experiment (shown in Fig. 7.30) validates the response speed of this control method.

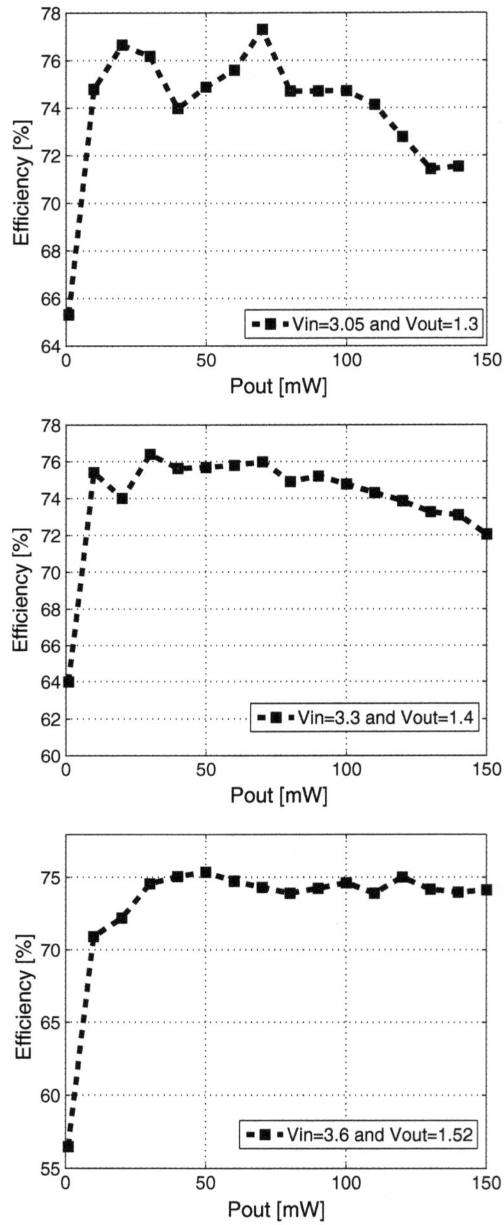

Fig. 7.26 Measurements of the efficiency in function of output power for 3 different input-output voltage combinations

Variable Input

Capacitive DC–DC converters are often penalized because of their limited voltage conversion flexibility compared to inductive DC–DC converters. As has been dis-

Table 7.3 Charge balance analysis

Topology	$Q_{out}/Q - \phi_1$	$Q_{out}/Q - \phi_2$	$Q_{in}/Q - \phi_1$	$Q_{in}/Q - \phi_2$
4/5	1	3/2	1	1
2/3	1	2	1	1
1/2	2	2	1	1

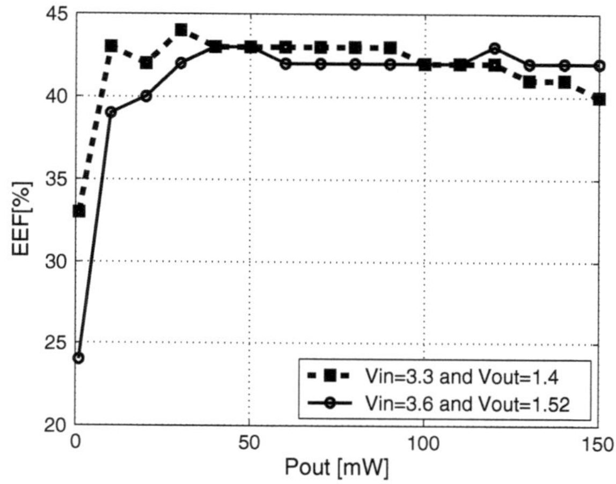

Fig. 7.27 Measurements of the Efficiency Enhancement Factor over the entire power range

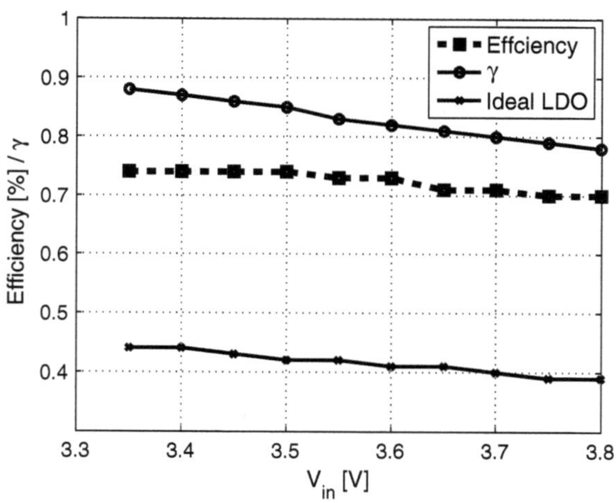

Fig. 7.28 Measurement of the efficiency for an input voltage range (3.35–3.8 V), 60 mA load and a fixed output voltage of 1.48 V

cussed in the previous sections: they have an optimal VCR for which they achieve high efficiency. In theory a shift towards lower VCR (and thus lower γ) will reduce the intrinsic maximum efficiency, while for inductive converters there is no fundamental impact of a change of VCR on the efficiency. In practice inductive converters also demonstrate a reduction in efficiency as a consequence of the change in duty

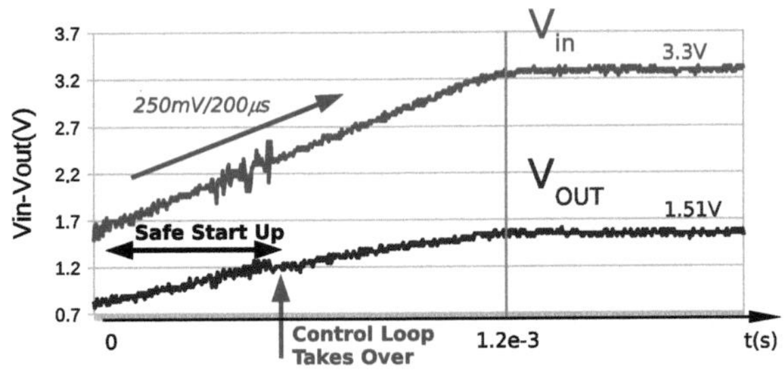

Fig. 7.29 Start-up behavior: when the input voltage ramps up, the output voltage is ramped up simultaneously by the start-up circuit. Until the output voltage is high enough and the control loop takes over

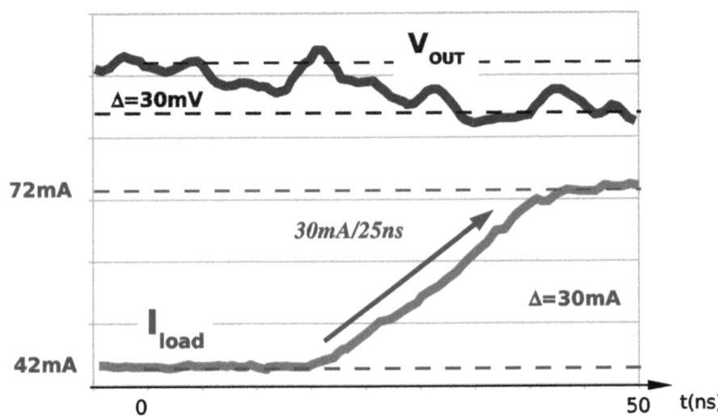

Fig. 7.30 Demonstration of the load regulation bandwidth. When a load step of 30 mA is applied with a rise time of 25 ns, the DC–DC converters remains stable and maintains control. (V_{in} 3.3 V V_{out} 1.4 V)

cycle and increased resistive losses when reducing the VCR. For capacitive DC–DC converters the increase in intrinsic losses is countered by a reduction in extrinsic losses since the switching frequency decreases to establish the drop in VCR. So that for capacitive DC–DC converters the system efficiency is less heavily impacted by a decrease in VCR than γ is. In a real-life application the converter is either supplied by a battery or a loosely regulated external supply, while the output voltage of the converter needs to be stable and constant at all times. In order to demonstrate the ability of this capacitive DC–DC converter to deal with input voltage variations (due to battery discharge for example) the following measurement is executed: In Fig. 7.28 the performance of the presented DC–DC converter is shown for a 60 mA load and an output voltage of 1.48 V. Although the intrinsic efficiency drops by 10 % over the input voltage range—the overall efficiency experiences a reduction of only 4 % and the efficiency remains much higher than the potential maximum efficiency of an ideal LDO.

Start Up

In Sect. 7.3.4 the voltage-domain stacking technique has been discussed. Obviously the devices in the circuit must be protected against voltages higher than the breakdown voltages of the technology. Higher voltages might endanger the reliability of the circuit through Hot-Carrier-Degradation and Negative-Bias Temperature Instability

(NBTI) Maricau and Gielen (2010). This can only be done by ensuring the voltage-domain boundary conditions during steady-state as well as during start-up. For this reason the following start-up circuit (shown in Fig. 7.23) is conceived. The start-up circuit clamps the output node and the input voltage node during start-up. This ensures appropriate voltage protection for the upper voltage domain. The logic supply V_r is ramped up together with the output voltage V_{out} because of the linear regulator and the control loop starts to operate once $V_r = 0.7\,\text{V}$. During a short time the start-up circuit and the control loop operate simultaneously, reinforcing each other. Once V_{out} reaches a reference voltage (1.2 V) the start-up circuit is deactivated. This procedure ensures the voltage boundary conditions of the lower voltage domain. The only devices that might be exposed to higher voltages across their terminals are the MIM capacitors, but these devices can withstand voltages above 10 V so that actually none of the nominal voltages are exceeded. Figure 7.29 shows a measurement of the start up behavior and clearly describes the above-stated behavior.

7.3.6 Conclusion

A fully integrated capacitive DC–DC converter has been presented which is depicted in Fig. 7.31. The converter achieves a peak efficiency of 77 % and an efficiency of 74 % for a maximum load of 150 mW. A multi-loop multi-phase control scheme is proposed, which enables multi-phase for medium power monolithic DC–DC converters without stability issues.

7.4 Phase-Handover Hysteretic Capacitive Converter with Feed-Forward Topology Control

7.4.1 Introduction

Recent work on monolithic capacitive DC–DC converters combines high power density with high efficiency. But next to power density and efficiency also output noise is an important steady-state performance metric. Multi-phase interleaving has grown to become a frequently proposed technique to reduce the ripple in capacitive DC–DC converters. Although it is a frequently proposed technique, the control of these interleaved converters is hardly discussed or even ignored. In Le et al. (2010) an open-loop converter is demonstrated, in Chang et al. (2010) a single-phase converter is presented while suggesting multi-phase interleaving as a potential noise-reduction technique. In Breussegem and Steyaert (2010) a hysteretic controller is presented to control a multi-phase interleaved converter, but the quality of the control relies heavily on the matching between a set of ten comparators and moreover the controller does not ensure equal current sharing between the converter cores. The latter becomes

Fig. 7.31 Micro-photograph of the chip: 10 sqmm including bond pads and ESD, 5 sqmm active part

even more important when moving towards higher power densities. A second recent evolution is the implementation of various topologies with different conversion ratios by means of a single switch-capacitor array Le et al. (2010); Breussegem and Steyaert (2010). None of these prototypes include an on-chip control loop that selects the appropriate topology.

7.4.2 Summary

A 200 mW monolithically integrated capacitive DC–DC converter in a 140 nm BCD SOI CMOS technology is presented. It combines three topologies in a single switch-capacitor core to bridge the gap between a battery voltage (3.6–2.5 V) curve at the input and a four-mode (1.2-1.4-1.6-1.8 V @max 100 mA) voltage-scaling power-management unit at the load side. To increase the power density a 2×5 interleaving multi-phase approach is used. Additionally, a Phase-Handover (PH) single-boundary hysteretic controller is proposed to improve the current sharing while compromising neither the power efficiency nor the control bandwidth. A Feed-Forward (F^2) technique is used for topology selection without interfering with the load control.

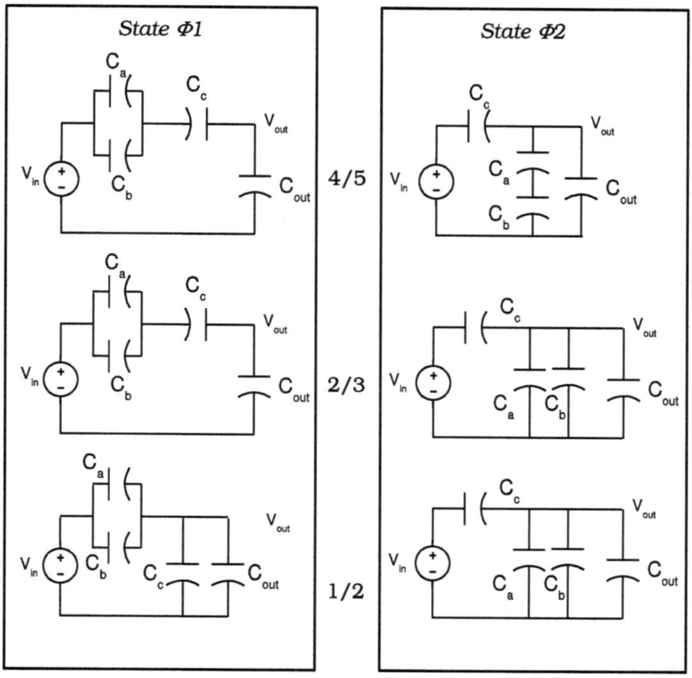

Fig. 7.32 Converter Topology: The switch-capacitor structure exists from fifteen switches and three capacitors, switch according to two states. The topology control activates the required configuration

Table 7.4 Charge balance analysis

Topology	Q_{out}/Q - ϕ_1	Q_{out}/Q - ϕ_2	Q_{in}/Q - ϕ_1	Q_{in}/Q- ϕ_2
4/5	$1/2 + 3/4$	$3/4 + 1/2$	$1/2 + 1/2$	$1/2 + 1/2$

7.4.3 Converter Structure

To cover the broad input-output range, three fractional topologies have been selected: 1/2-2/3-4/5. The topologies are configured by means of a 15-transistor switch array. Thanks to the common states in both the 2/3 -4/5 and the 1/2 -2/3 topology, the switch overhead is reduced. Moreover by sizing the capacitors $C_a = C_b = C_c/2$ the charge transfer is optimized (output impedance reduced) as discussed in Sect. 3.1 and this has been implemented for all three configurations. In Fig. 7.32 the converter topologies are represented for the three distinct topologies in both conversion states.

Charge balance analysis, based on the technique introduced in Sect. 2.2, identifies an unbalanced charge transfer at the output for the 2/3 and 4/5 topology. The results of this analysis are demonstrated in Table 7.3.

Table 7.5 Performance overview

Process	ABCD9 - 140 nm SOI
$iVCR$	$1/2, 2/3, 4/5$
# Phases	5×2 times Multi-phase
V_{in}	3.6–2.5 V
V_{out} Regulated	Continuously between 1.2–1.8 V
C_{fly}	10 nF MOS-cap
C_{out}	5 nF MOS-cap
$f_{sw,max}$	80 MHz
η_{max}	74 %
$P_{out,max}$	200 mW
$\eta_{P_{out,max}}$	70 %
EEF_{max}	47 %
$I_{in,noload}$	$2\,mA$
Load Reg.max	−700 mΩ

For example for topology 4/5: if an amount of charge Q is transferred during state Φ_1 by capacitor C_c, then both C_a and C_b each transfer $Q/2$ during that state. As a consequence the amount of charge that flows into the output during state Φ_1 is Q (only capacitor C_c is connected to the output) and an equal amount is taken from the input by the parallel connection of C_a and C_b.

For state Φ_2 the charge transfer by the capacitors is equal as in state Φ_1 if the converter resides in steady state. But now the amount of charge that flows into the output during state Φ_1 is the charge from both C_c and C_a. This results in $Q + Q/2$ while an amount Q is taken from the input by C_c. Obviously the charge dump in the output capacitor during state Φ_1 differs from Φ_2. While the charge taken from the input is equal during both states. This is clearly indicated in Table 7.3.

This unbalanced charge transfer[3] results in an unequal ripple voltage at the output node. The charge transfer can be equalized by fragmenting the converter by a factor 2, paralleling both fragments and activating the fragments in anti-phase. For example for the 4/5 topology this results in one converter transferring $Q_{out}/Q = 1/2$, while the other transfers $Q_{out}/Q = 3/4$. The next state, the fragments switch roles and the first transfers $Q_{out}/Q = 3/4$ while the second transfers $Q_{out}/Q = 1/2$. This balanced case is represented in Table 7.4. As long as this fragmentation has no impact on the capacitor's quality, this can be done without area nor efficiency penalty.

The converter core, depicted in Fig. 7.33, is the structure that is responsible for the voltage conversion. It consists of the switch-capacitor array, the power trains that drive the switches, a topology decoder and the non-overlap circuitry. In fact it groups

[3] Unbalanced charge transfer denotes that the absolute amount of charge transferred to the load during state ϕ_1 is not equal to the charge transferred to the load during state ϕ_2.

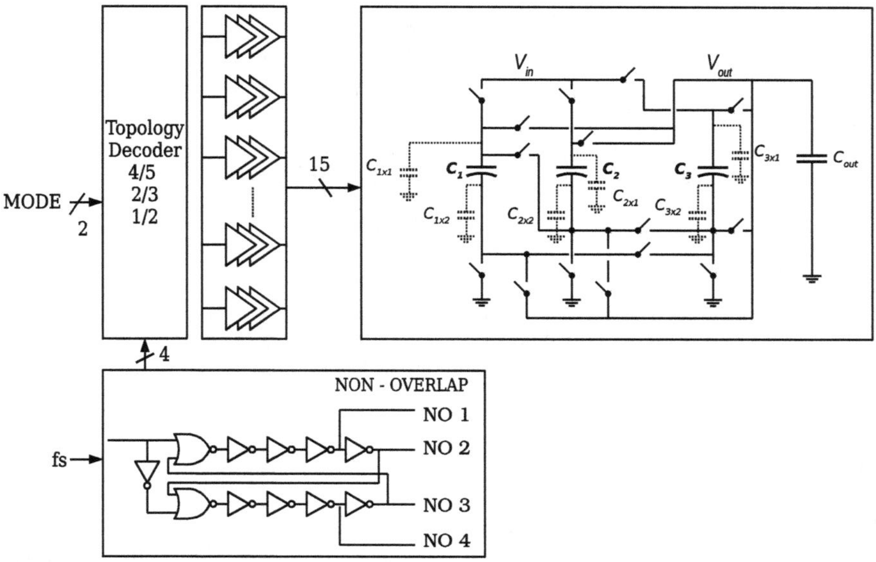

Fig. 7.33 Converter Core Structure: the switch-capacitor array, the power trains that drive the switches, a topology decoder and the non-overlap circuitry

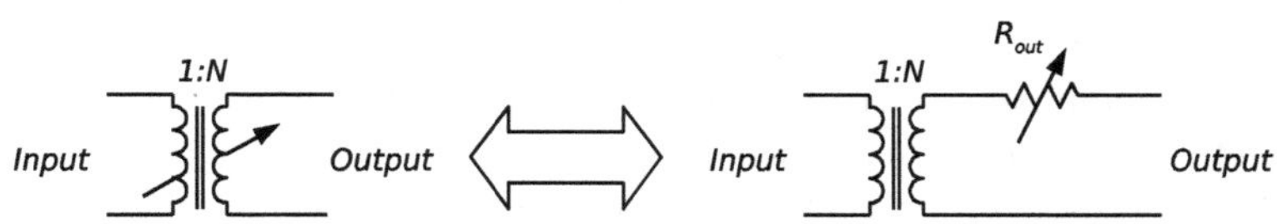

Fig. 7.34 Comparison of an ideal DC–DC converter with the model of a capacitive DC–DC converter

all the time-critical circuitry, so that the global routing of the IC has a minimum impact on the converter's conversion performance.

7.4.4 System

To obtain a capacitive DC–DC converter that mimics the behavior of an ideal DC–DC converter a reconfigurable switch-capacitor array, controlled by a dual control loop has been implemented. This control loop comprises a first controller which modifies the output impedance by hysteretic frequency control to obtain a constant output voltage disregarding the load and a second controller to reconfigure the converter's topology to match the optimum iVCR (Fig. 7.34).

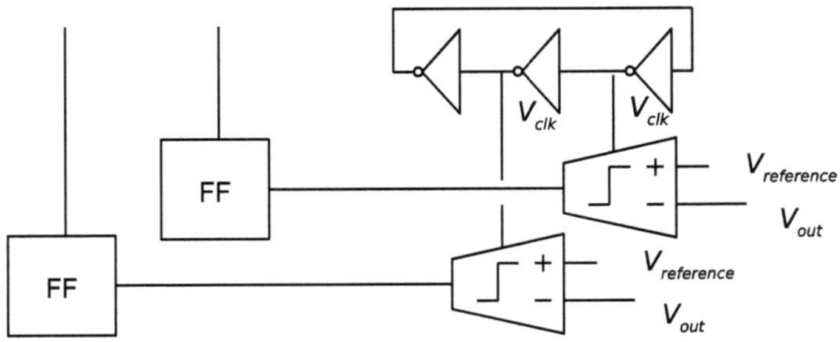

Fig. 7.35 The multi-loop multi-phase hysteretic control loop: Every core has a dedicated hysteretic control loop. These loops are shifted out of phase by synchronizing the comparators in the loop with different clock signals derived from a multi-tap oscillator

Load Control

The most straightforward technique to perform load control in a capacitive DC–DC converter is by modulating the switching frequency of the DC–DC converter and as such changing the output impedance of the capacitive DC–DC converter. By means of this technique the output impedance of the converter is masked and a constant output voltage is sustained even when the load current changes.

This switching frequency modulation can be implemented by means of an analog control loop comprising an error amplifier and a voltage-controlled oscillator or by a hysteretic comparator-based control loop. In Sect. 5.3 the advantages of hysteretic control with respect to the conventional analog control techniques are discussed in detail. Without doubt the digital and low-power nature of the hysteretic control method are considered the most prominent advantages.

But in state-of-the-art capacitive DC–DC converters multi-phase interleaving, introduced in Sect. 4.4, has become one of the predominant techniques to reduce the converter's output capacitor without compromising the output ripple. This results in a more compact design and thus a higher power density. Therefore it is equally important to reconcile multi-phase interleaving with hysteretic control techniques. It has been illustrated in Fig. 7.35 how to implement a hysteretic multi-loop multi-phase switching scheme. For every converter core a hysteretic control loop is foreseen. By synchronizing the comparators of every loop with a distinct clock signal-phase shifted with respect to the other clock signals the converter cores are activated according to a multi-phase pattern.

This control loop, although a low-power and high-bandwidth digital solution, has two important drawbacks. The first issue is the requirement of tight matching of the comparators. A difference in comparator offset voltage results in erroneous operation of the control loop and an increase in converter ripple. This is demonstrated in Fig. 7.36, in that particular case the second comparator is subject to a significant mismatch resulting in a virtual shift in reference voltage. Therefore this particular comparator can not detect a hysteretic boundary violation. The missed boundary violation is detected by the next activated comparator but this results in larger ripple.

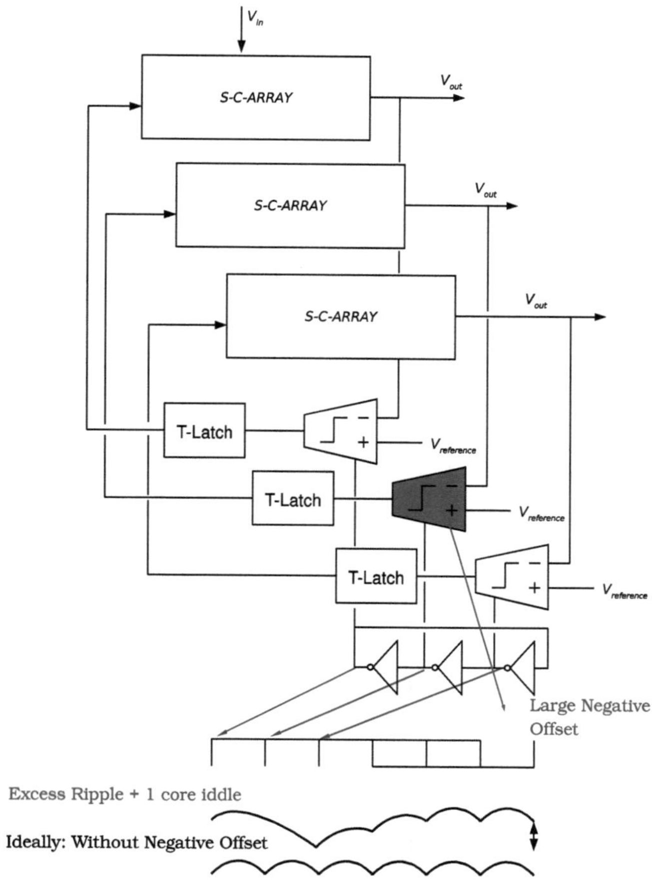

Fig. 7.36 Effect of comparator offset in the multi-loop approach

Since the control signal of all comparators is common, the comparator offset can not be corrected externally by changing the control voltage level. This would influence the control level for the remaining comparators as well. Implementing an individual offset correction for every comparator introduces a significant overhead, which is unwanted.

Next to the offset-prone characteristics of this control technique, this multi-loop technique does not ensure equal current sharing between the converter cores. Equal current sharing is very well appreciated in the design of DC–DC converters and thus must be included. Phase unbalance results in thermal runaway and overstress of parts of the converter and is thus highly undesirable especially when migrating towards higher power densities.

These obstacles can be resolved by implementing the control loop as is demonstrated in Fig. 7.37. To omit the sensitivity to comparator offset, a single comparator is used instead of an N_{MP}-dimensional comparator array. Eventual comparator offset

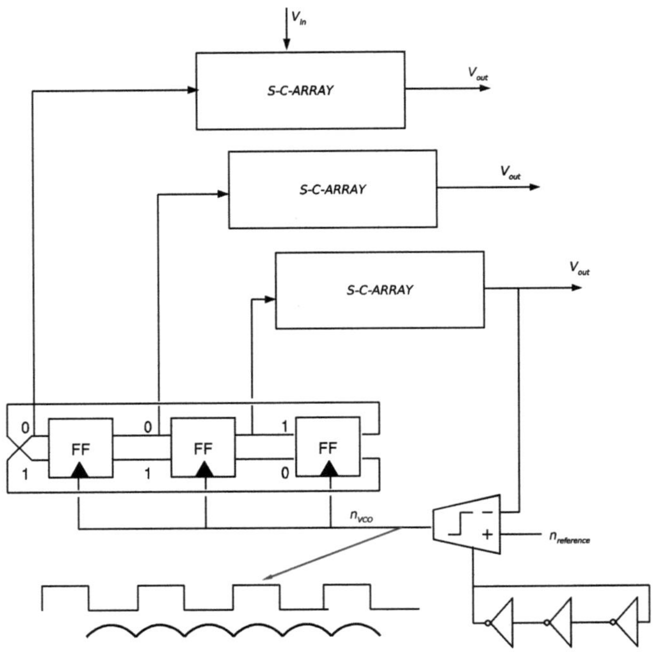

Fig. 7.37 Single-loop hysteretic controller

now translates into an offset of the control signal and can be corrected easily since the comparator is not sharing this signal.

In the multi-loop approach, the phase separation is ensured by activating the comparators at distinct time intervals. In case of the single-loop approach, phase separation is endorsed by a shift register. This shift register is connected as such that each clock pulse toggles a single bit in the register. The clock of this shift register is generated by the comparator. For committing to an equivalent switching frequency as in the multi-loop approach, the comparator runs at an N_{MP} higher frequency than the comparators in the multi-loop approach. But due to the equal reduction in hardware, both approaches result in equal power consumption.

The differential nature of the shift register ensures a deterministic bit pattern to be loaded: [0 0 1] and [1 1 0] in the complementary string. This is necessary to invoke single core activation. The shift register also introduces a fixed order of activation of the cores irrespective of the load or time interval in which a boundary violation is detected.

In Fig. 7.38 a typical waveform pattern in the single-loop hysteretic controller is demonstrated. The comparator is operated as a clocked boundary-violation detector. When $V_{out} < V_{reference}$ the VCO clock is passed through to the shift phase-handover block and signal $COMP$ is generated for which the rising edge of $COMP$ corresponds to the boundary violation. If no boundary violation is detected, the comparator gates the VCO signal. The $COMP$ signal is used as a clock for the phase-handover

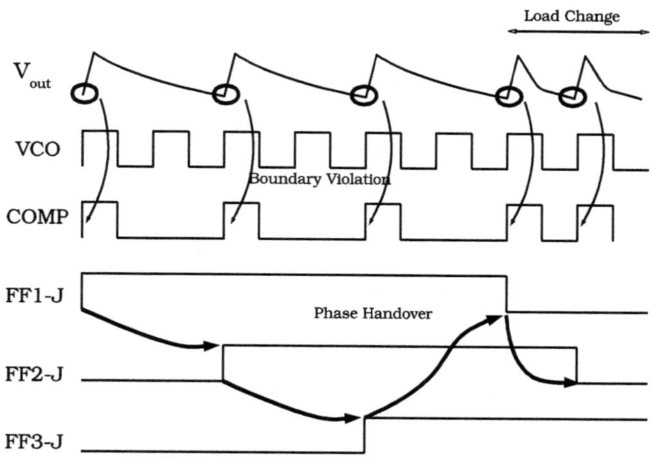

Fig. 7.38 A single-loop hysteretic controller's waveform

block and this block shifts around the bitsequence initiating one core after the other but at the same time masks the falling edges of the $COMP$ signal. This would lead to erroneous activation of the converter cores. It is demonstrated in Fig. 7.38 that one cores is activated after the other in an order defined by the shiftregister's structure. After three phase hand-over actions, the situation is sketched that a change in load occurs. This results in an increase in effective switching frequency to anticipate a voltage drop at the output.

This method provides a low power method to reconcile multi-phase interleaving with hysteretic control including current-sharing properties and resilience against comparator offset. Additionally the output of the comparator can be bypassed by an external clock or on-chip VCO to provide an open loop testing mode.

Line Control

In addition to the load control, a topology-selection method is proposed. This method maps the topology of the capacitive converter on the actual input voltage and the desired output voltage represented by the voltage reference level. The mapping is executed as follows:

$$V_{in} \times N_{i+1} < V_{reference} <= V_{in} \times N_i \tag{7.4}$$

$$\rightarrow VCR = N_i, \tag{7.5}$$

$$\left(\frac{1}{2}, \frac{2}{3}, \frac{4}{5}\right) \bigcap N \tag{7.6}$$

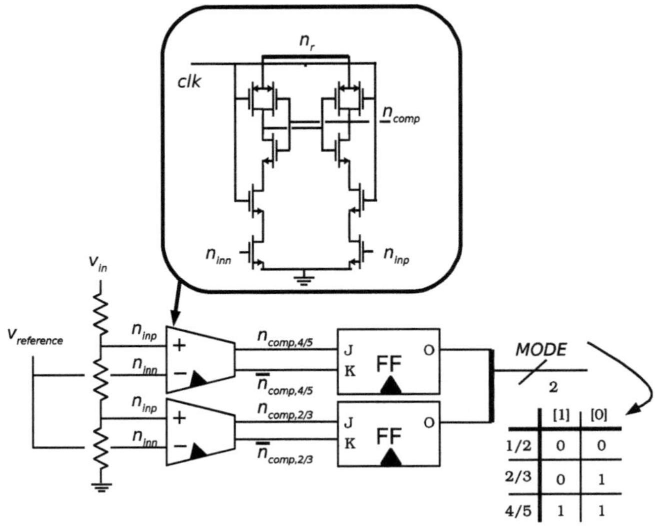

Fig. 7.39 The topology mapping by means of a clocked-comparator-based look-up table

Section 2.3 discusses how capacitive DC–DC converters have an upper limit for the VCR that can be achieved: the iVCR. But additionally the highest efficiency is achieved close to the iVCR. By adding additional topologies and reconfiguring, the range of VCR's for which a high efficiency can be attained, is extended.

The above formulation neglects the output impedance of the converter, which renders it in practice impossible to execute a capacitive conversion close to the iVCR. Therefore an output impedance penalty factor is introduced. This penalty factor $K_{penalty}$ decreases the acceptance slope and a new topology selection algorithm is defined:

$$K_{penalty} V_{in} \times N_{i+1} < V_{reference} <= K_{penalty} V_{in} \times N_i \tag{7.7}$$

This mechanism has been implemented by means of an Feed-Forward operating flash-type ADC. This Feed-Forward controller is depicted in Fig. 7.39 and uses two comparators and register to determine the topology. The resistors are sized according to the given acceptance slopes including the output impedance penalty factor.

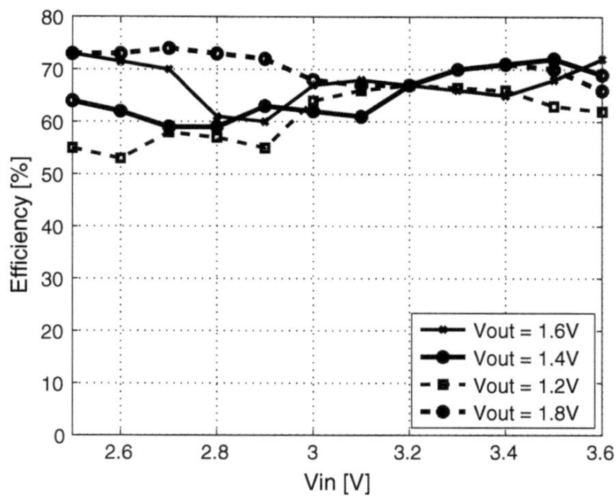

Fig. 7.40 Efficiency in function of the input voltage with respect to the four output voltages

7.4.5 Measurement Results

Efficiency Measurement

Figure 7.40 demonstrates the efficiency for the four main output voltage nodes over the input voltage range (2.5–3.6 V) at a constant load of 100 mA. A maximum efficiency of 74 % is attained for the conversion of 2.7–1.8 V.

Figure 7.41 shows the efficiency curve of the converter measured as a function of the load current for a conversion from an input of 3.6 V to the four output voltages. The maximum efficiency of 74 % is achieved given a 80 mA load and a 1.6 V output voltage. It has been demonstrated before that for frequency-modulated capacitive DC–DC converters the efficiency is relatively flat as a function of the load current. The efficiency roll of at the lower end of the load range (<10 mA) is invoked by the increasing effect of the static power consumption in the on-chip oscillator. This circuit consumes a considerable amount of power with respect to the output power at the loads (1–2 mA). The comparators and digital logic demonstrate a power consumption smaller than 100 μA.

Efficiency Enhancement

In order to normalize the converter's efficiency η_{sw} improvement with respect to a linear regulator η_{LR} the Efficiency Enhancement Factor EEF, is used.[4] In Fig. 7.42 the EEF is plotted of the converter over the range of interest. For the main part of conversion pairs the EEF lies within 10 % and 47 %, resulting in significant reduction of battery power in a real-life application.

[4] The EEF is introduced and motivated in Sect. 1.3.

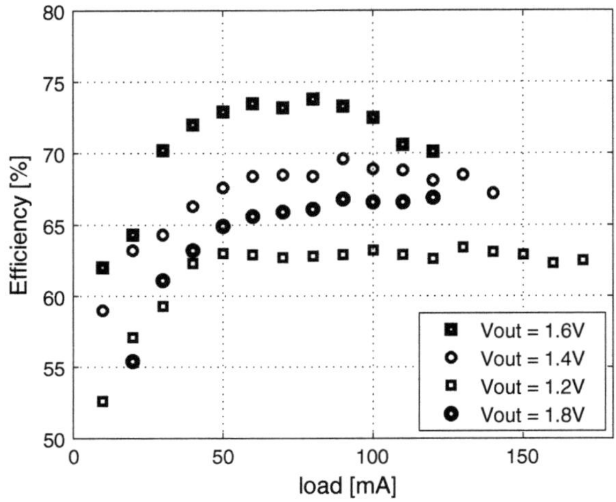

Fig. 7.41 Efficiency in function of load current for a conversion from 3.6 to 1.2 V, 1.4 V, 1.6 V, 1.8 V

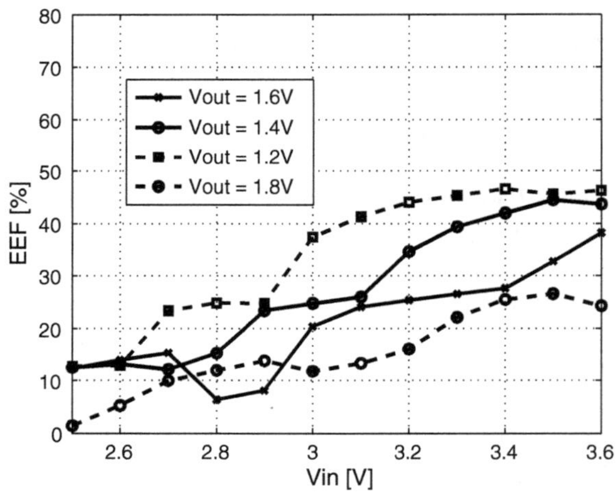

Fig. 7.42 Efficiency Enhancement function of the input voltage with respect to the four output voltages

Line Regulation

In Fig. 7.43 an input voltage slope from 3.6 to 2.5 V is applied, while fixing the output at 1.6 V and supplying 80 mA to a resistive load. The upper slope is the input voltage, the constant line is the output voltage. At an input voltage of 2.7 V a change in topology occurs without a significant impact on the output voltage. For the lower part of the input range a 2/3-topology is selected, which is automatically reconfigured to a 1/2-topology when this threshold is violated. During the change in input voltage, the hysteretic single boundary control matches the output minimum voltage with the 1.6 V reference voltage.

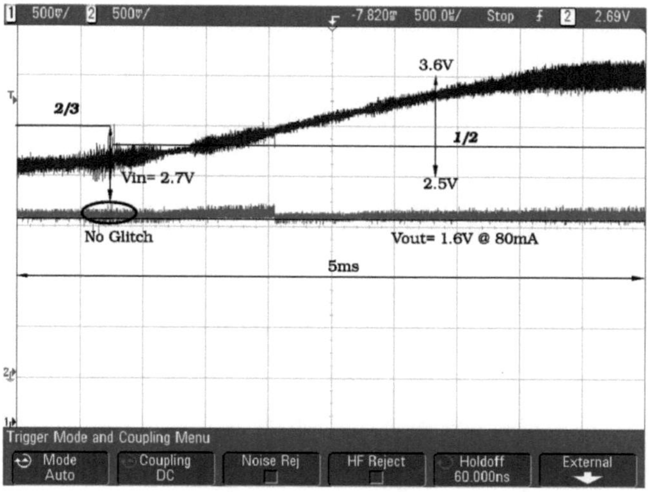

Fig. 7.43 Line regulation for a full swing on the input terminal and a constant output voltage of 1.6 V at a load of 80 mA

Voltage Scaling

Figure 7.44 demonstrates the voltage-scaling properties for a 2.5 V input voltage and a maximum output voltage step at the output (1.2–1.8 V). Within 100 ns the output changes from the minimum to the maximum voltage. This variation is accomplished by the F^2 loop changing the topology from 2/3 to 4/5 in combination with the hysteretic loop adapting the switching frequency to the new load condition (20 Ohm load at 1.8 V instead of at 1.2 V). The ripple of the converter is smaller than 100 mV and falls within the 10 % requirement.

Load Regulation

It is theoretically determined in Sect. 5.3 that hysteretic controllers posses the ability to correct the output voltage of a capacitive converter within one cycle. Therefore hysteretic single-boundary control anticipates on sudden load variations without a significant under- nor overshoot. Figure 7.45 demonstrates the capacitive converter output subject to a load drop of 100 mA. No significant transient behavior is observed, the effective switching frequency is adapted instantly as predicted by theory. The capacitive converter in this particular case operates in a 1/2-topology while generating a 1.2 V output from a 3 V supply.

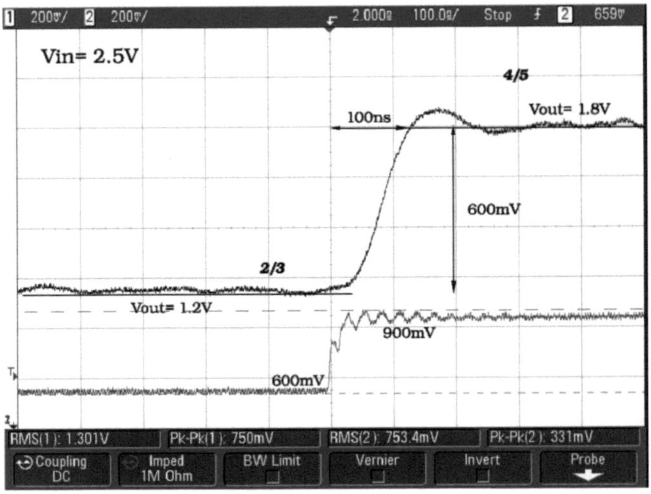

Fig. 7.44 Voltage scaling for a full swing on the output terminal and a constant input voltage of 2.5 V at a load of 80 mA

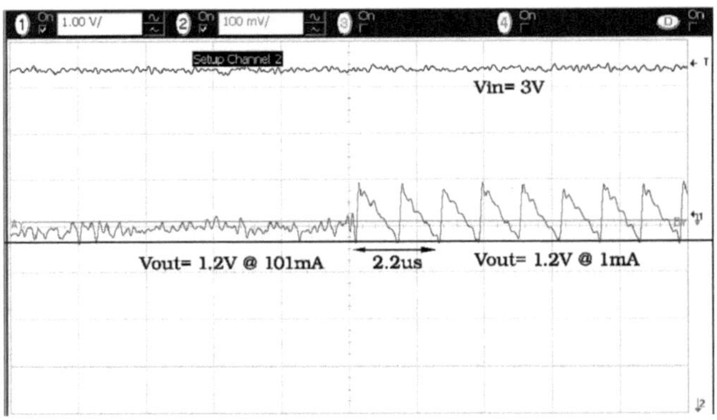

Fig. 7.45 Line regulation for a full swing on the input terminal and a constant output voltage of 1.6 V at a load of 80 mA

7.4.6 Conclusion

This final prototype demonstrates a capacitive DC–DC converter that unifies most of the techniques developed in this work. First the multi-phase interleaving by fragmentation is further exploited by addressing the fragmentation for achieving a balanced charge transfer to the output capacitor. Secondly, the multi-loop control technique for multi-phase interleaving is replaced by the more robust single-loop alternative. Finally a fully functional on-chip topology selection circuit is proposed.

7.5 Conclusion

This chapter gave a detailed discussion of four capacitive DC–DC converter prototypes. It became clear that capacitive converters are much more than a power stage. The peripheral control circuitry plays an at least equally important role. The prototypes demonstrate how hysteretic control alleviates the power constraints in the control loop by implementing the complexity by means of digital circuits. The use of a hysteretic technique introduces a lot of potential with respect to a straightforward analog solution. But this can only be exploited when unification between hysteretic control and multi-phase interleaving is achieved.

Chapter 8
Conclusions

This book focuses on the capacitive DC–DC converter as an alternative for the conventional inductive type of DC–DC converters. In fact the field of capacitive DC–DC conversion is little explored compared to the extensive literature that exists on inductive type of converters. This not without reason of course, since capacitive DC–DC converters typically require a larger number of components than inductive converters. Moreover the capacitive nature prohibits lossless conversion, in contrast to inductive converters which ideally achieve 100 % efficiency at a finite switching frequency. Clearly there are enough reasons to prefer an inductive DC–DC converter over a capacitive type.

But the evolution from discrete type of converters towards fully integrated converters has altered the rules of the game significantly. In this new context the number of components is no longer an issue since these components are monolithically integrated. And in an integrated case the inductive converters are cut short by the poor quality of the integrated inductors, the key-components in the design. Therefore the intuitive preference for inductive converters does not hold anymore. Moreover, integrated capacitors—crucial for the operation of the capacitive converters—are native devices in CMOS technology and can be constructed at high quality.

All these aspects turn the capacitive DC–DC converter in an appropriate candidate to become the next generation of DC–DC converters for fully integrated solutions. The following paragraphs summarize the main conclusions drawn from this work and highlight the novel techniques developed or explored.

8.1 Need for On-Chip DC–DC Conversion

The main drivers behind the need for on-chip DC–DC converters are the proliferation of power loss in state-of-the-art digital SoC's, the voltage gap paradigm and the energy gap. For dealing with the increase in power loss and the energy gap a number of power-management techniques have been consolidated. The most successfully

T. Van Breussegem and M. Steyaert, *CMOS Integrated Capacitive DC–DC Converters*, Analog Circuits and Signal Processing, DOI: 10.1007/978-1-4614-4280-6_8,

used techniques are dynamic voltage scaling and body biasing. In the state-of-the-art implementations these power-management techniques still rely on either discrete-type converters or monolithic linear regulators to manage the supply voltage. The first solution results in a large volume due to increased number of chips and the linear regulator provides a compact but low-efficient way to control the supply voltage.

The voltage gap poses an even more stringent problem: how to power circuits in small-feature technologies with battery voltages much higher than the breakdown voltage of these technologies? It is clear that on-chip DC–DC converters not only enable interfacing between the supply and the low-voltage circuits but also reduce the power consumption by using smart power management.

The beneficial nature of CMOS with respect to the capacitors turns CMOS in a very favorable platform to develop capacitive-type converters. Especially since the alternative—inductive-type converters—failed to demonstrate large improvement with respect to a linear regulator. The maximum EEF—which benchmarks the DC–DC converter with respect to an ideal linear regulator—demonstrated in Wens and Steyaert (2011) is 23 %. The capacitive converters, demonstrated in this work, present EEF values up to 47 %. This shows that, at least in terms of EEF, the capacitive DC–DC converter outperforms the inductive type of DC–DC converter.

It must be asserted that the use of capacitive DC–DC converters is subject to a number of obstacles. First, the switching nature and the impulse-like charge transfer—typical for switched-capacitor structures—introduces significant amount of noise in the system. In this work a technique, multi-phase interleaving, is successfully adopted to reduce the switching noise. Secondly, literature demonstrates few control techniques to manage the capacitive DC–DC converters.

8.2 DC–DC Converter Types for Fully Integrated Power Management

The selection of the appropriate DC–DC converter for a certain application is twofold. First one has to choose between a linear or a switched-mode DC–DC converter. Next, if a switched mode topology ispreferred, selection of the appropriate topology: inductive versus capacitive DC–DC converter is required.

The choice between a linear regulator and a switched-mode DC–DC converter is quite straightforward. That is the choice between efficiency and cost. A linear regulator can be build with a minimum of components and requires little chip-area. But inherently the maximum efficiency is constrained by the VCR. In case the VCR is close to 1: efficiency can be high. When the VCR is low, the linear regulator will dissipate considerable amount of power with respect to the output power.

If efficiency is of high importance a switched-mode converter appears to be a viable alternative, since efficiency is ideally independent of VCR and this at the cost of chip area. Trustworthy comparison between an inductive and capacitive solution for on-chip DC–DC conversion must be based on a thorough assessment of the technology and the application dependent converter characteristics. This book provides a number of stead-state circuit-evaluation techniques (f.e. the Output Impedance Model and the OIB-technique) to predict the circuit performance and compare it to the outcome of steady-state models of inductive DC–DC converters.

Although the strong dependency on technology characteristics the following guidelines can be given:

- Due to the entanglement of VCR and topology in case of capacitive DC–DC converters. The latter demonstrate excellent results in case that an application requires voltage conversion with a fixed VCR.
- If for capacitive DC–DC converters, a large number of topologies are required. The benefit of an additional topology must be assessed on the circuit level. While an additional topology might improve the ideal efficiency of the VCR-range, this might compromise the eventual performance due to the additional parasitics. Which does not necessarily mean that inductive converters perform better in this case.
- Capacitive converter structure is often more complex than a simple inductive converter (f.e. the buck converter) but this provides more opportunities to evade certain issues. For example by addressing voltage-domain stack in capacitive DC–DC converters instead of switch stacking in buck converters.

These findings are supported by the current state-of-the-art which is illustrated in Sect. 1.4. This section discusses a number of trendsetting prototypes and gives an overview of all recent prototypes based on the prior-defined FoMs.

8.3 Noise Reduction in Fully Integrated DC–DC Converters

As mentioned throughout the previous section: noise is one of the most critical characteristics in an electronic circuit. Especially in large SoC's mitigation of supply noise is to be taken very seriously. The switched nature of both inductive and capacitive DC–DC converters turns these circuits into one of the main aggressors in a mixed-signal design. Hence noise reduction is next to efficiency the major concern when designing a switched-mode DC–DC converter.

Multi-phase interleaving is a technique which is adopted from the area of inductive converters, but it was until recently that this technique's potential has been recognized as valuable for capacitive DC–DC converters. Moreover, multi-phase interleaving demonstrates even more potential when applied to capacitive DC–DC converters than to inductive converters.

Multi-phase interleaving in capacitive converters has the following benefits:

1. Output noise is reduced by smearing out the charge transfer in time
2. Power density is increased since the output capacitor can be omitted

But to ensure no penalty in efficiency, the following must be true:

1. Introducing multi-phase interleaving has negligible penalty on the control power
2. Fragmentation of the switches and passives has negligible effect on their quality

It has been demonstrated in Chap. 6 that the latter assumptions hold and capacitive converters in contrast to inductive DC–DC converters benefit significantly from this technique. Throughout this work, multi-phase interleaving is promoted as a major driver behind the acceptance of capacitive DC–DC converters in power management.

8.4 Control of Fully Integrated DC–DC Converters

The advent of fully integrated DC–DC converters requires new control techniques to support the specific needs of fully integrated power-management solutions. First the control techniques require to be low power with respect to the minimum output power of the DC–DC converter. Secondly, the high switching frequencies—relative to the of-the-shelf DC–DC converters—demand for alternative control methods. And finally the noise-reduction techniques, which are proposed in Chap. 4, have to be supported by these control methods.

Scientific literature on the early capacitive DC–DC converters reveals little information on how to build a control loop for these devices. Either the control loop is built based on techniques used in inductive DC–DC converters or the control loop is omitted entirely. Therefore, control-loop design and implementation has been one of the cornerstones in this work and this based on the conviction that the best possible control method is the one that embraces the fundamental nature of the capacitive DC–DC converter.

The first prototype in Chap. 7 demonstrates how by lead-compensation, the highly-variable pole of the voltage doubler is omitted. Moreover, the multi-phase nature of the DC–DC converter is supported by a Cascaded Clock-Divider circuit. But obviously, the constant power consumption of the OTA puts heavy constraints on the power budget of the control loop. Therefore a highly digital approach is favored in the second prototype, since the power consumption of such a loop can be much lower. And by adding a capacitance modulation feature to the hysteretic controller, both ripple and load regulation are improved. A third prototype aimed for unifying the multi-phase switching technique with the hysteretic control method. This led to the development of a multi-loop multi-phase hysteretic control technique. A state-space analysis demonstrated the robustness and high bandwidth of this technique and clearly proved the close relationship between control of the converter and the fundamental nature of the capacitive DC–DC converter.

Finally the fourth prototype included a Feed-Forward loop to control the topology selection of a three-topology converter stage and an improved single-loop multi-phase hysteretic controller for capacitive DC–DC converters. The latter perfectionizes the multi-phase controller proposed by means of the third prototype.

References

M. Alimadadi, S. Sheikhaei, G. Lemieux, S. Mirabbasi, P. Palmer, A 3 ghz switching dc–dc converter using clock-tree charge-recycling in 90nm cmos with integrated output filter, in *Solid-State Circuits Conference, 2007. ISSCC 2007. Digest of Technical Papers. IEEE International*, pp. 532–620, 2007

M. Alon, E. Horowitz, Integrated regulation for energy-efficient digital circuits. IEEE J. Solid State Circ. **43**(8), 1795–1807 (2008)

P. Andreani, S. Mattisson, On the use of mos varactors in rf vcos. IEEE J. Solid State Circ. **35**(6), 905–910 (2000)

M. Badaroglu, G. Vander Plas, P. Wambacq, L. Balasubramanian, K. Tiri, I. Verbauwhede, S. Donnay, G.G.E. Gielen, H.J. De Man, Digital circuit capacitance and switching analysis for ground bounce in ics with a high-ohmic substrate. IEEE J. Solid State Circ. **39**(7), 1119–1130 (2004)

H.J. Bergveld, R. Karadi, K. Nowak, An inductive down converter system-in-package for integrated power management in battery-powered applications, in *IEEE Proceedings of the Power Electronics Specialists Conference*, pp. 3335–3341, 2008

H.J. Bergveld, K. Nowak, R. Karadi, S. Iochem, J. Ferreira, S. Ledain, E. Pieraerts, M. Pommier, A 65-nm-cmos 100-mhz 87%-efficient dc–dc down converter based on dual-die system-in-package integration, in *IEEE Energy Conversion Congress and Exposition, 2009. ECCE 2009*, pp. 3698–3705, 2009

Bsim4.3.0 mosfet model, Manual

J.N. Burghartz, M. Soyuer, K.A. Jenkins, M. Kies, M. Dolan, K.J. Stein, J. Malinowski, D.L. Harame, Integrated rf components in a sige bipolar technology. IEEE J. Solid State Circ. **32**(9), 1440–1445 (1997)

T. Burd, T. Pering, A. Stratakos, R. Brodersen, A dynamic voltage scaled microprocessor system, in *Solid-State Circuits Conference, 2000. Digest of Technical Papers. ISSCC. 2000 IEEE International*, pp. 294–295, 466, 2000

S. Ben-Yaakov, A. Kushnerov, Algebraic foundation of self adjusting switched capacitors converters, in *IEEE Energy Conversion Congress and Exposition, 2009. ECCE 2009*, pp. 1582–1589, 2009

L. Chang, D.J. Frank, R.K. Montoye, S.J. Koester, B.L. Ji, P.W. Coteus, R.H. Dennard, W. Haensch, Practical strategies for power-efficient computing technologies. Proc. IEEE, **98**(2), 215–236 (2010)

L. Chang, R.K. Montoye, B.L. Ji, A.J. Weger, K.G. Stawiasz, R.H. Dennard, A fully-integrated switched-capacitor 2:1 voltage converter with regulation capability and 90% efficiency at 2.3a/mm2, in *IEEE Proceedings of the Symposium on VLSI Circuits*, pp. 55–56, 2010

T. Van Breussegem and M. Steyaert, *CMOS Integrated Capacitive DC-DC Converters*, Analog Circuits and Signal Processing, DOI: 10.1007/978-1-4614-4280-6,

C.Y. Cheng, K.N. Leung, Y.K. Sun, P.Y. Or, Design of a low-voltage CMOS charge pump, in *IEEE Proceedings of the International Symposium on Electronic Design, Test and Applications*, pp. 342–345, 2008

R.C. Chu, R.E. Simons, M.J. Ellsworth, R.R. Schmidt, V.V. Cozzolino, Review of cooling technologies for computer products. Device Mater. Reliab. IEEE Trans. **4**(4), 568–585 (2004)

L.T. Clark, E. Hoffman, M. Schaecher, M. Biyani, D. Roberts, Y. Liao, A scalable performance 32 b microprocessor, in *Solid-State Circuits Conference, 2001. Digest of Technical Papers. ISSCC. 2001 IEEE International*, pp. 230–231, 2001

J. Craninckx, M. Steyaert, A 1.8-ghz low-phase-noise spiral-lc cmos vco, in *VLSI Circuits, 1996. Digest of Technical Papers, 1996 Symposium on*, pp. 30–31, 1996

S. Cuk, Modeling, Analysis, and Design of Switching Converters, Ph.D. thesis, California Institute, 1976

G. Delagi, Harnessing technology to advance the next-generation mobile user-experience, in *Solid-State Circuits Conference Digest of Technical Papers (ISSCC), 2010 IEEE International*, pp. 18–24, 2010

J.F. Dickson, On-chip high-voltage generation in MNOS integrated circuits using an improved voltage multiplier technique. IEEE J. Solid State Circ. **11**(3), 374–378 (1976)

R. Dorf, *Moderm Control Systems* (Prentice Hall, Upper Saddle River, 1995)

M.J. Ellsworth, Chip power density and module cooling technology projections for the current decade, in *Thermal and Thermomechanical Phenomena in Electronic Systems, 2004. ITHERM '04. The Ninth Intersociety Conference on*, vol. 2, pp. 707–708, 2004

E. Evennat, P. Lorrain, The Cockcroft–Walton voltage multiplying circuit. Sci. Inst. **24**(3), 221–226 (1953)

P. Favrat, P. Deval, M.J. Declercq, A high-efficiency CMOS voltage doubler. IEEE J. Solid State Circ. **33**(3), 410–416 (1998)

B.R. Gregoire, A compact switched-capacitor regulated charge pump power supply. IEEE J. Solid State Circ. **41**(8), 1944–1953 (2006)

P. Hazucha, G. Schrom, J. Hahn, B.A. Bloechel, P. Hack, G.E. Dermer, S. Narendra, D. Gardner, T. Karnik, V. De, S. Borkar, A 233-mhz 80–87% efficient four-phase dc–dc converter utilizing air-core inductors on package. IEEE J. Solid State Circ. **40**(4), 838–845 (2005)

The international technology roadmap for semiconductors 2009, public.itrs.net., 2009

G.C. Verghese, J.G. Kassakian, M.F. Schlecht, *Principles Power Electronics* (Addison-Wesley, Reading, 1991)

P.T. Krein, J. Bentsman, R.M. Bass, B.L. Lesieutre, On the use of averaging for the analysis of power electronic systems. IEEE Trans. Power Electron. **5**(2), 182–190 (1990)

P.T. Krein, *Elements of Power Electronics* (Oxford University Press, Oxford, 1998)

J. Kwong, Y.K. Ramadass, N. Verma, A.P Chandrakasan, A 65 nm Sub-V_t microcontroller with integrated SRAM and switched capacitor dc–dc converter. IEEE J. Solid State Circ. **44**(1), 115–126, 2009

K.L. Sai, P.K.T. Mok, K.N. Leung, A low-dropout regulator for soc with q-reduction. IEEE J. Solid State Circ. **42**(3), 658–664 (2007)

H. Le, M. Seeman, S. Sanders, V. Sathe, S. Naffziger, E. Alon, A 32 nm fully integrated reconfigurable switched-capacitor dc–dc converter delivering 0.55 W/mm^2 at 81% efficiency, in *IEEE Proceedings of the International Solid State Circuits Conference*, pp. 210–211, 2010

J.-Y. Lee, S.-E. Kim, S.-J. Song, J.-K. Kim, K. Sunyoung, H.-J. Yoo, A regulated charge pump with small ripple voltage and fast start-up. IEEE J. Solid State Circ. **41**(2), 425–432 (2006)

J. Lee, G. Hatcher, L. Vandenberghe, C.-K. Ken Yang, Evaluation of fully-integrated switching regulators for cmos process technologies. IEEE Trans. VLSI Syst. **15**(9), 1017–1027 (2007)

M.S. Makowski, D. Maksimovic, Performance limits of switched-capacitor dc–dc converters, in *IEEE Proceedings of Power Electronics Specialists Conference*, pp. 1215–1221, 1995

E. Maricau, G. Gielen, Efficient variability-aware nbti and hot carrier circuit reliability analysis. IEEE Trans. Comput. Aided Des. Integr. Circ. Syst. **29**(12), (2010)

Max1682, switched capacitor voltage doubler, Datasheet

Max16945, inverting charge pump in sot23 for emi-sensitive automotive applications, Datasheet
L. Mensi, L. Colalongo, A. Richelli, Z.M. Kovacs-Vajna, A new integrated charge pump architecture using dynamic biasing of pass transistors, in *Proceedings of the European Solid-State Circuits Conference*, pp. 85–88, 2005
H. Meyvaert, T. Van Breussegem, M. Steyaert, A monolithic 0.77w/mm^2 power dense capacitive dc–dc step-down converter in 90nm bulk cmos, in *Proceedings of the ESSCIRC*, 2011
S. Minkyu, I. Chowdhury, D. Ma, A.P. Brokaw, Adaptive step-down switched-capacitor power converter with z-domain observation-based line-load regulation, in *IEEE Proceedings of the Symposium on VLSI Circuits*, pp. 134–135, 2008
Mp1530, triple output step-up plus linear regulators for tft bias, Datasheet
V.W. Ng, M.D. Seeman, S.R. Sanders, High-efficiency, 12v-to-1.5v dc–dc converter realized with switched-capacitor architecture, in *IEEE Proceedings of the Symposium on VLSI Circuits*, pp. 168–169, 2009
J. Ni, Z. Hong, B.Y. Liu, Improved on-chip components for integrated dc–dc converters in 0.13 um cmos, in *Proceedings of ESSCIRC, 2009. ESSCIRC '09*, pp. 448–451, 2009
W.P. Robbins, N. Mohan, T.M. Undeland, *Power Electronics* (Wiley, New York, 2002)
R. Perigny, U. Moon, G. Temes, Area efficient CMOS charge pump circuits. Proc. IEEE Int. Symp. Circ. Syst. **1**, 492–495 (2001)
J.M. Rabaey, A. Chandrakasan, B. Nikolic, *Digital Integrated Circuits* (Prentice Hall, Upper Saddle River, 1996)
A.P. Chandrakasan, Y.K. Ramadass, Voltage scalable switched capacitor dc–dc converter for ultra-low-power on-chip applications, in *IEEE Proceedings of the Power Electronics Specialists Conference*, pp. 2353–2359, 2007
Y.K. Ramadass, A.A. Fayed, A.P. Chandrakasan, A fully-integrated switched-capacitor step-down dc–dc converter with digital capacitance modulation in 45 nm cmos. IEEE J. Solid State Circ. **45**(12), 2557–2565 (2010)
B. Razavi, R.-H. Yan, K.F. Lee, Impact of distributed gate resistance on the performance of mos devices. IEEE Trans. Circ. Syst. I Fundam. Theory Appl. **41**(11), 750–754 (1994)
A. Richelli, L. Colalongo, M. Quarantelli, M. Carmina, Z.M. Kovacs-Vajna, A fully integrated inductor-based 1.8–6-V step-up converter. IEEE J. Solid State Circ. **39**(1), 242–245 (2004)
J. Rius, M. Meijer, A model for on-chip decoupling capacitor effectiveness including gate leakage effects, in *IEEE 13th Topical Meeting on Electrical Performance of Electronic Packaging*, pp. 299–302, 2004
D. Maksimovic, R.W. Erickson, *Fundamentals of Power Electronics* (Kluwer Academic Publishers, Norwell, 2001)
L.G. Salit, Y.I. Ismail, Fast hysteretic control of on-chip multi-phase switched-capacitor dc–dc converters, in *IEEE International Symposium on Circuits and Systems (ISCAS)*, pp. 2561–2564, 2011
L.G. Salit, R. Jain, A novel control technique to eliminate output-voltage-ripple in switched-capacitor dc–dc converters, in *IEEE International Symposium on Circuits and Systems (ISCAS)*, pp. 825–828, 2011
H. Samavati, A. Hajimiri, A.R. Shahani, G.N. Nasserbakht, T.H. Lee, Fractal capacitors. IEEE J. Solid State Circ. **33**(12), 2035–2041 (1998)
W. Sansen, *Analog Design Essentials*, The Springer International Series in Engineering and Computer Science, 2006
S. Sawant, U. Desai, G. Shamanna, L. Sharma, M. Ranade, A. Agarwal, S. Dakshinamurthy, R. Narayanan, A 32nm westmere-ex xeon; enterprise processor, in *Solid-State Circuits Conference Digest of Technical Papers (ISSCC), 2011 IEEE International*, pp. 74–75, 2011
M. Seeman, S. Sanders, Analysis and optimization of switched-capacitor DC/DC converters. IEEE Trans. Power Electron. **23**(2), 841–851 (2008)
M.D. Seeman, S.R. Sanders, J.M. Rabaey, An ultra-low-power power management ic for energy-scavenged wireless sensor nodes, in *IEEE Power Electronics Specialists Conference*, pp. 925–931, 2008

M.D. Seeman, S.R. Sanders, J.M. Rabaey, An ultra-low-power power management ic for energy-scavenged wireless sensor nodes, in *IEEE Proceedings of the Power Electronics Specialists Conference*, pp. 925–931, 2008

M.D. Seeman, V.W. Ng, H.-P. Le, M. John, E. Alon, S.R. Sanders, A comparative analysis of switched-capacitor and inductor-based dc–dc conversion technologies, in *IEEE 12th Workshop on Control and Modeling for Power Electronics (COMPEL)*, pp. 1–7, 2010

D. Senderowicz, G. Nicollini, S. Pernici, A. Nagari, P. Confalonieri, C. Dallavalle, Low-voltage double-sampled delta; converters. IEEE J. Solid State Circ. **32**(12), 1907–1919 (1997)

B.J. Sheu, C. Hu, P.K. Ko, F.-C. Hsu, Source-and-drain series resistance of ldd mosfet's. IEEE Electron Device Lett. **5**(9), 365–367 (1984)

D. Somasekhar, B. Srinivasan, G. Pandya, F. Hamzaoglu, M. Khellah, T. Karnik, K. Zhang, Multi-phase 1ghz voltage doubler charge-pump in 32nm logic process, in *IEEE Proceedings of the Symposium on VLSI Circuits*, pp. 196–197, 2009

D. Somasekhar, B. Srinivasan, G. Pandya, F. Hamzaoglu, M. Khellah, T. Karnik, K. Zhang, Multi-phase 1 ghz voltage doubler charge pump in 32 nm logic process. IEEE J. Solid State Circ. **45**(4), 751–758 (2010)

J.A. Starzyk, Y.-W. Jan, F. Qiu, A dc–dc charge pump design based on voltage doublers. IEEE Trans. Circ. Syst. I Fundam. Theory Appl. **48**(3), 350–359 (2001)

J.T. Stauth, S.R. Sanders, Optimum biasing for parallel hybrid switching-linear regulators. IEEE Trans. Power Electron. **22**(5), 1978–1985 (2007)

J.T. Stauth, S.R. Sanders, Power supply rejection for rf amplifiers: theory and measurements. IEEE Trans. Microw. Theory Tech. **55**(10), 2043–2052 (2007)

M. Steyaert, P. Vancorenland, Cmos: a paradigm for low power wireless?, in *IEEE Proceedings of the Design Automation Conference*, pp. 836–841, 2002

F. Su, W.H. Ki, C.Y. Tsui, An SC voltage regulator with novel area-efficient continuous output regulation by dual-branch interleaving control scheme, *IEEE Proceedings of the Symposium on VLSI Circuits*, pp. 136–137, 2008

L. Su, D. Ma, A. Brokaw, Design and analysis of monolithic step-down SC power converter with subthreshold DPWM control for self-powered wireless sensors. IEEE Trans. Circ. Syst. I **57**(1), 280–290 (2010)

J. Tschanz, J. Kao, S. Narendra, R. Nair, D. Antoniadis, A. Chandrakasan, V. De, Adaptive body bias for reducing impacts of die-to-die and within-die parameter variations on microprocessor frequency and leakage, in *Solid-State Circuits Conference, 2002. Digest of Technical Papers. ISSCC. 2002 IEEE International*, vol. 1, pp. 422–478, 2002

Y. Tsividis, Principles of operation and analysis of switched-capacitor circuits. Proc. IEEE **71**(8), 926–940 (1983)

T. Van Breussegem, M. Steyaert, A 82% efficiency 0.5% ripple 16-phase fully integrated capacitive voltage doubler, in *IEEE Proceedings of the Symposium on VLSI Circuits*, pp. 198–199, 2009

T. Van Breussegem, M. Steyaert, A fully integrated 74% efficiency 3.6v to 1.5v 150mw capacitive point-of-load dc/dc-converter, in *Proceedings of the ESSCIRC*, 2010

T. Van Breussegem, M. Steyaert, A fully integrated gearbox capacitive dc/dc-converter in 90nm cmos: optimization, control and measurements, in *IEEE 12th Workshop on Control and Modeling for Power Electronics*, pp. 1–5, 2010

T.M. Van Breussegem, M.S.J. Steyaert, Compact low swing gearbox-type integrated capacitive dc/dc converter. Electron. Lett. **46**(13), 892–894 (2010)

T.M. Van Breussegem, M.S.J. Steyaert, Monolithic capacitive DC–DC converter with single Boundary–Multiphase control and voltage domain stacking in 90 nm CMOS. IEEE J. Solid State Circ. (2011)

V. Von Kaenel, P. Macken, M.G.R. Degrauwe, A voltage reduction technique for battery-operated systems. IEEE J. Solid State Circ. **25**(5), 1136–1140 (1990)

M. Wens, M. Steyaert, A fully-integrated 0.18μm CMOS dc–dc step-up converter using a bondwire spiral inductor, in *IEEE Proceedings of the European Solid State Circuits Conference*, pp. 268–271, 2007

M. Wens, K. Cornelissens, M. Steyaert, A fully-integrated 130nm CMOS dc–dc step-down converter, regulated by a constant on/off-time control system, in *IEEE Proceedings of the European Solid State Circuits Conference*, pp. 62–65, 2008

M. Wens, M.S.J. Steyaert, A fully integrated cmos 800-mw four-phase semiconstant on/off-time step-down converter. IEEE Trans. Power Electron. **26**(2), 326–333 (2011)

M. Wens, M. Steyaert, *Monolithic Inductive DC–DC Converters* (Springer, New York, 2011)

J. Wibben, R. Harjani, A high-efficiency dc/dc converter using 2 nH integrated inductors. IEEE J. Solid State Circ. **43**(4), 844–854 (2008)

Curse of dimensionality, Site

W.C. Wu, R.M. Bass, Analysis of charge pumps using charge balance, in *IEEE Proceedings of the Power Electronics Specialists Conference*, vol. 3, 2000

C.-H. Wu, C.-L. Chen, A low-ripple charge pump with continuous pumping current control, in *IEEE Proceedings of the Midwest Symposium on Circuits and Systems*, pp. 722–725, 2008

J. Zhang, H.S.H. Chung, W.-L. Lo, S.Y. Hui, A.K.-M. Wu, Implementation of a decoupled optimization technique for design of switching regulators using genetic algorithms. IEEE Trans. Power Electron. **16**(6), 752–763 (2001)

M. Zhang, N. Llaser, Optimization design of the Dickson charge pump circuit with a resistive load, in *Proceedings of the International Symposium on Circuits and Systems*, vol. 5, 2004

J. Zhang, H.S.-H. Chung, A.W.-L. Lo, T. Huang, Extended ant colony optimization algorithm for power electronic circuit design. IEEE Trans. Power Electron. **24**(1), 147–162 (2009)

C. Zheng, D. Ma, A 10mhz 92.1%-efficiency green-mode automatic reconfigurable switching converter with adaptively compensated single-bound hysteresis control, in *Solid-State Circuits Conference Digest of Technical Papers (ISSCC), 2010 IEEE International*, pp. 204–205, 2010

G. Zhu, A. Ioinovici, Switched-capacitor power supplies: dc voltage ratio, efficency, ripple, regulation, pp. 553–556, 1996

MIX
Papier aus verantwortungsvollen Quellen
Paper from responsible sources
FSC® C105338

If you have any concerns about our products,
you can contact us on
ProductSafety@springernature.com

In case Publisher is established outside the EU,
the EU authorized representative is:
Springer Nature Customer Service Center GmbH
Europaplatz 3, 69115 Heidelberg, Germany

Printed by Libri Plureos GmbH
in Hamburg, Germany